Bob Grossblatt's Guide to Creative Circuit Design

Writing a book is a mind-consuming business, and it's easy to forget about the day-to-day things in life. So, for looking after the house, the food, and especially for looking after me, this book is dedicated To My Wife Barbara.

It'll be nice to get reacquainted.

Bob Grossblatt's Guide to Creative Circuit Design

Robert Grossblatt

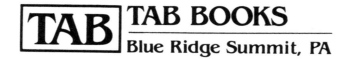

TAB BOOKS
Blue Ridge Summit, PA

FIRST EDITION
FIRST PRINTING

© 1992 by **TAB Books**.
TAB Books is a division of McGraw-Hill, Inc.

Library of Congress Cataloging-in-Publication Data

Grossblatt, Robert.
   Bob Grossblatt's guide to creative circuit design / by Robert
Grossblatt.
     p.   cm.
   Includes index.
   ISBN 0-8306-7610-4   ISBN 0-8306-3610-2 (pbk.)
   1. Electronic circuit design. I. Title.
TK7867.G74 1992
621.381′5—dc20   91-34395
                                            CIP

TAB Books offers software for sale. For information and a catalog, please
contact TAB Software Department, Blue Ridge Summit, PA 17294-0850.

Acquisitions Editor: Roland S. Phelps
Technical Editor: Kay Maloney
Director of Production: Katherine G. Brown
Book Design: Jaclyn J. Boone
Managing Editor: Sandra L. Johnson
Paperbound Cover Photo: Susan Riley, Harrisonburg, VA   EL1

Contents

Acknowledgments

I've wanted to do a book like this for a long time. Over the years I've dealt with a lot of designers, both neophytes and experts(?), and I've found that the biggest problem people had was that they just didn't know how to think logically. People in the business for a long time usually knew a lot of stuff, but there was never a guarantee that they could be systematic about applying what they knew.

The people who deserve the most credit in this book are people who took the time to sit down and write me a letter telling me about the projects they were working on. Some people deserve special attention as well as some companies that made it a lot easier to get through the pages.

In no particular order:

- My editor, Roland Phelps, who put up with lots of delays and earned my undying gratitude by not trying to make me feel bad.
- My friend Dennis Jump of the Great Softwestern Company, whose parts and symbol libraries made doing the artwork, if not actually enjoyable, at least tolerable.
- All the people at Autodesk. Without AutoCAD, it would have taken me so long to produce the artwork that the parts used in the circuits would have been obsolete before I finished the book. AutoCAD will always be the standard against which other CAD software is judged—and Autosketch is still the best deal around.

It takes a lot of time, work, and energy to produce a book like this, and I couldn't have done it without the understanding of some of my other clients. They extended deadlines and did without so I could spend their time on my work.

Even though all of this book came out of my brain, other people put some of the stuff into it. They all know who they are, and I hope they don't get too bent because they're not being thanked individually. If any of them are offended, I hereby apologize. If you let me know, I promise to make it up to you.

God give me strength.

Introduction

One of the greatest things about writing regularly for a magazine is reading the mail you get—well, reading some of it, anyway. Some of the mail is absolutely unbelievable! The letters that aren't written by the strange people not only give me ideas on subjects to write about, but introduce me to some terrific people as well.

As you can imagine, lots of people get in touch with me simply because they're too lazy to think for themselves. I can't tell you how many letters I've gotten that start out "Could you design something for me that . . . ."

Now I don't mind helping out, but there are limits. The letters I like are the ones from people who have spent lots of bench time on an idea and are stuck with some kind of glitch in the design. If somebody has already paid dues, I'll do whatever I can to help out.

Over the years, I've noticed that while different circuit problems have different solutions, there are really only three basic reasons why problems show up in the first place:

- A lack of practical experience or electronic understanding
- A mechanical error in wiring—miscounting IC pins, misreading resistor color codes, and so on
- Bad work habits at the bench or failure to follow the right rules for developing a design

When I first started out in this business, I would have laid serious money that the biggest cause of screwups in circuit development was that people just didn't know enough to get the job done. And even if they did, the board didn't work because of a slipup in the wiring.

After lots of experience, I'm here to tell you that I was completely wrong. Believe it or not, the main cause of prototype circuits that don't do anything except drain batteries is the method used to design the circuit in the first place.

Most people have problems in electronic design because they're less than systematic in how they go about designing the circuit. This is an important point, and I want to make sure you understand it.

Taking a project from a good idea to a working reality is a complex business. Most people sit down for about 15 minutes with a pen and paper to sketch the idea and then start working things out at the bench. This method might be the natural product of a shortage of time or an abundance of enthusiasm, but it's also a good way to help guarantee failure.

No matter what you're trying to build—from an electronic toothbrush to a nitron beam (and who remembers that?)—the method is always the same. A certain progression of steps should be followed if you want to be successful. As a matter of fact, some really down-and-dirty market research has shown me that a lot more projects get started than get finished. The more often you beat your head against a wall, the more tempting it becomes to quit and start something else.

Books will teach you just about anything you want to learn about electronics, but I've yet to find one that shows you how to go about designing your own circuits. Don't listen to the people who want to pump your head full of technical knowledge without spending time on explaining why things are like that or what to do with what they're teaching. Brain space is valuable real estate, and it's stupid to fill it with stuff you can look up in a book. You can get facts anywhere—from a book, someone who knows, even signs on a bus—so cramming them into your head is only a good thing if you want to make a career out of TV game shows. Besides, how many sets of matched luggage do you really need?

Knowing about stuff is a good thing, but knowing what to do with stuff is even better. The Forty-Fourth Law of Life and Design, with apologies to Fats Waller (or someone) is:

It ain't what you use, it's the way that you use it.

Method is a lot more important than materials.

As you go through this book, you'll see that how you do things is as critical a part of the job as the components used in the design itself.

One of the great truths is that nothing winds up the way it was planned, and that's true here as well. By the time you get to the end of a design job, there's usually little resemblance between what you had in mind when you started and what you have on the bench when you're finished. Ideas change, goals change; that's just the way things work—always. As you work on something, you're often hit with practical considerations that you couldn't see when you were still in the planning stage. The road from idea to actuality is not a straight line.

A smart designer has to keep an open mind while working because it's really impossible to foresee every consequence of every design decision. Ideas that seem terrific when you begin can turn out to be ridiculous when you really get into working on the circuit. The best designs are the result of creative evolution at the bench. Locking yourself into an initial idea is a bad idea. You have to be open to all the wild possibilities.

Before we get started on our trip to designland, there's one more thing you should keep uppermost in your mind. No matter what you're working on or even how you go about designing it, all your bench time is wasted unless you see the job through to the end. The one really unshakable principle is to finish what you start. Nobody wants to hear about what you didn't do, and saying stuff like "I could have" is the ultimate wimp-out.

To say "I could have done it, but I didn't know how," "I could have made it work, but I couldn't get the parts," or "I would have done it, but it blew up" is a cop-out. The bottom line is that you didn't finish it.

Of all the ingredients that go into a completed design, the most important one is perseverance. Never lose sight of the fact that the reason you started something in the beginning was to have it finished at the end. Don't let yourself be overwhelmed by design complications or defeated by self-doubt. Your job, your only job, is to make the design work.

And so now the Fifth Law of Life and Design:

Elegance comes later.

Things can always be made better; the goal for version 1 is just to get it working, not to get it perfect. Getting through the initial design, as I said before, is just about impossible unless you follow the rules.

By the time you finish this book, you'll know how to analyze an idea and make your bench time as efficient as possible. It's never easy to build a working prototype; but if you understand the method, it's much, much easier.

1
The basic idea
What it's all about

As you go through life, you read all sorts of things, pick up all sorts of weird information; and for some reason the darndest things get stuck in your brain forever. There's no way to figure out why or how it happens, but I'll bet every one of you has a small collection of useless bits of esoterica buried between your ears.

When I was 12 years old or so, I read one of those "Believe It or Not" columns in a newspaper, and the lead item was the existence of a two-headed cow. Now this in itself was pretty strange, but what really stuck in my mind was what happened when one of the heads wanted to walk to the left and the other wanted to go to the right. The animal would stand still for a second and then fall down as its two front legs went in different directions.

The reason I thought of this is simply that lots of people who get a notion to design a particular circuit never get their idea off the ground because they get completely overwhelmed by the complexity and size of things in general. The more they think about the problem, the less they're able to see how to get started. Now this isn't so much of a problem if all you're trying to put together is a light dimmer or some other simple circuit; but as the size of the idea gets larger, so too does the confusion.

The only real difference between the design of a single-component light dimmer and a 60-gazillion-component space shuttle is how much you'll have to do to get it done. Admittedly you'll have certain other things to worry about with the space shuttle (how to get it out of the basement when you're done, etc.), but the basic approach to designing it is the same as you'd use for a light dimmer. Seriously.

Now lots of people get an idea in their heads and immediately sit down at the bench to work it out. Fifteen minutes later they're facing a mass of wires and components and, if they're lucky, something interesting will happen when they finally turn the device on. It might not do what they had in mind, but at least it'll do something.

This is "Mission Impossible" stuff. Cannibalize a broken radio and turn it into a radar set. Perfect stereo sound from a single mike hidden in the false bottom of a whiskey bottle. Strike a match on a bar of soap. Stuff like that.

It's nice to see this sort of junk, but let's get real. There's no more chance of this happening than there is of finding intelligent life on Pluto. The only matches you'll be able to strike on a bar of soap are the ones the fish-faced people of Neptune used to make out of unobtainium.

In the real world where, unfortunately, most of us spend our days, this sort of gee-whiz stuff is impossible. Translating an idea to a working circuit means a lot of work and a lot of time. Unless you go about it the right way, you'll spend a whole lot more time and energy than you have to with absolutely no guarantee that you'll wind up with anything worthwhile to show for it.

Unless you understand the right way to design something, you're not going to be successful. I know that this sounds like I'm beating you over the head, but there's no getting around the fact that it's true. While you'll find lots of interesting circuit ideas and neat circuit tricks in the pages of this book, the method used to arrive at the design is a whole lot more important than the design itself.

Thinking, the first step

It may be hard to believe, and for some people hard to do, but the first step in any design is thinking about what you want to do. No matter what you have in mind, your first thoughts should be nothing more than a rough idea of the project you want to build. First impressions, after all, are nothing more than first impressions. Even if you have the world's most orderly mind, there's no way an initial idea can begin to cover all the details of the project. Even the "Mission Impossible" people think about the stuff they do.

Spending pure brain time on an idea means thinking about it and not thinking about it. I know this sounds weird, but don't forget that there are two ways to work out the details of a problem—consciously and unconsciously. I can't tell you how many times I've been stuck on something only to have the answer staring at me as soon as I opened my eyes in the morning. Consciously thinking about a problem is good for laying out the parameters, but if you can't work the problem out while it's uppermost in your mind, leave it to your subconscious.

It's taken me a long time to admit to myself that my subconscious is a better problem solver than I am. After all, you like to think that, if nothing else, you have some control over yourself. I've finally gotten to the point where I know when trying to consciously solve a problem is only going to be a waste of time. The bottom line is that if I understand what the problem is and keep coming up with the same wrong answers, I'll get up from the bench and go get a pastrami sandwich.

Because I usually work alone (at least in the planning stage of a design project), I've come to think of this schizophrenic view of my mind as a private kind of brainstorming. When you start a project as a member of a team, the early stages are loaded with what are euphemistically referred to as brainstorming sessions. This is when several people sit around a table in a locked room and try to outdo each other with ideas—occasional good ones and lots of bad ones. When you work on your own, the only kind of brainstorming you can do is with your subconscious.

This is an important part of the design process because there's no way you can work out a complete idea with 5 minutes or so of quick thought. Thinking and sleeping on an idea can only be to your benefit because the easiest mistakes to correct are the ones you haven't made yet. From the moment you come up with the initial idea, you should be writing everything down on paper—not just any scrap of paper but on the first page of a new notebook that you'll be using throughout the various stages of the project, from the first vague idea, through interim circuit developments, to the schematic of the completed design.

The notebook you're starting is probably the most important part of any project, and it never ceases to amaze me that so few people understand this concept. Remember that one of the reasons you're starting the project is that you can't find, afford, or use any commercial product. What you're planning to develop is something that's unique and, for better or worse, you're going to be the world's leading authority (the sole authority, for that matter) on the final product.

The notes you keep are the only notes there ever will be; and if you don't get in the habit of writing every single thing down, right from the beginning, you'll wind up shortchanging yourself later on when repair time comes or you want to move on to the next version.

Keeping a notebook is an essential part of the design process, no matter what it is you're working on—from electronics, to software development, to fixing the bathroom plumbing. A friend of mine is the head of a design team for Hughes, and one of his biggest complaints about new employees is that a lot of them have never been trained to keep accurate and complete records of the work they do. It became so big a deal that he created a corporate procedure detailing the need for keeping notes and outlining the form in which notes had to be kept. Individual notebooks were treated as corporate documents and were reviewed monthly. Anybody not keeping records of his or her work was warned the first time and canned the second time. Serious stuff.

Outlining, the second step

Thinking about a project and writing down the results of that thought in your new notebook are well and good, but sooner or later you're going to want to get started on the design. There's no hard-and-fast rule about knowing when it's time to start the design but, for me at least, I can tell because nothing really new gets added to the notes I've been keeping and I find myself glancing more and more often at the workbench.

Unfortunately, the desire to get your hands dirty has to be put off for a while longer because, at this stage of the game, you're still a ways removed from breadboards, wires, and all the rest of the fun stuff. There's still more paperwork to do.

As a result of all the thinking and notetaking you've done, you now have a good idea of what you want to do. By going through the notes you've written down, you should be able to clearly describe what you expect to have finished when you finally walk away from the bench and call in your family and friends so they can stare uncomprehendingly at the results of your labors. You may not know exactly how you're going to accomplish what you want to do (and at this stage of the game it would be unusual if you did), but you should know what kind of circuit you're going to design.

The next step in design is to formalize the results of all your thinking and note-taking by drawing up a list of design criteria. This list is just a formal, numbered list of the requirements that the proposed circuit has to satisfy. It should contain everything from how you want the circuit to be powered, to what you want it to do, and even include considerations about how it should be built and housed.

Out there in the real world, where new projects are backed by good things like real budgets filled with real money and bad things like real deadlines enforced by real bureaucrats, a list of design criteria is sometimes called the design specifications. However you want to refer to it, you won't find a design project anywhere that doesn't have as its foundation a set of design specs. The only differences between the list you put together and the one they put together are the number of people involved, the amount of time it took, and finally the undoubtedly huge wad of cash that was spent.

I'll tell you—from personal experience—that you're much, much better off. The more people you have involved in a design project, the longer it takes to get anything done and the more brain damage is generated along the way. There's an awful lot to be said for working alone.

Put some serious thought into what kind of things to include on your final list of design criteria because these specs are the foundation of all the work you're going to do. When I'm working on a project, I take a long time to compile my list; and once it's done, I nail the list to the wall over the bench so I'm forced to look at it whenever I'm working at the bench.

There's a good reason for this. Once you actually start creating circuitry, it's really easy to get so involved in the details that you lose sight of the overall goals of the project. Having the original specs staring you in the face is a good way to be constantly reminded of what you're aiming at. Even though the specs aren't written in stone, once you've arrived at your final list you should think twice about changing it.

The further along you are in a project, the more difficult and more expensive it is to change things. When the breadboard starts to fill up, it's a real pain in the neck to change things around. Sometimes even a good idea is just too much of a hassle to implement.

Fifteen or so minutes of thought when you're working with pen and paper can save you hours and days of laying out wire and components, to say nothing of making the final version of the circuit the best it can possibly be.

Planning, the third step

After having wrung the last dregs of good ideas out of your head, you now have a written goal: the list of design criteria. Even though you're no doubt aching to sit down at the bench and start the age-old design ritual of blowing things up, I'm sorry to tell you that there's still more paperwork waiting to be done.

Everything you've done so far has been aimed at figuring out what you want to do. The next step is figuring out how it's all going to work. You've got to analyze the list of design criteria and come up with a block diagram that breaks the project down into its functional parts and shows how they're all connected.

This part of the overall design procedure can be either complex or trivial, depending on what kind of project you have in mind. If your goal is to build a one- or two-component amplifier, you'll be able to get through this part of the job in about 15 minutes. Something like a microprocessor-based controller, however, is going to take you a lot more time. Whatever project you have in mind, the time it takes to come up with a block diagram can save you hours and days of wasted bench time later on.

A block diagram is an overall view of the completed circuit and is probably the most important part of the paperwork generated for the project. It breaks up even the world's most complex project into human-sized pieces and is an invaluable troubleshooting guide when the last bit of work is finally done and, as usually is the case, nothing works.

The real benefit of a block diagram in the initial design stages of a circuit is that it gives you a handle on the work you have to do. As a matter of fact, if you're planning to design something of any complexity whatever, the task is just about impossible without working out the block diagram before you sit down at the bench.

When we work on specific projects later in the book, you'll see how invaluable a block diagram can be. The boxes and arrows that outline the final project show you how many individual circuits you'll have to create and give you a good idea of the amount of time you'll have to spend on each one of them.

There's a more subtle, and even more important, reason for drawing a block diagram. While it's easy to see how having the skeleton of the circuit in front of you serves as a guide for the work you have to do, it's also easy to overlook the fact that designing from a block diagram almost forces you to concentrate on one part of the design at a time. Wearing these psychological blinders is a necessity when you're working on a large, complex project. There's no way you're going to be able to make something that works if your mind wanders from place to place. Think of one thing at a time—and only one.

The first step in translating your list of design criteria into a block diagram is to determine what sections you need and then to see how they're connected. First draw the boxes, then add the arrows.

Go through your design criteria and identify the individual parts of the circuit. The most obvious one, regardless of the nature of the final project, is the power supply. The power supply can be anything from a battery and capacitor to a full-blown, multivoltage, switching regulator that has to deliver 15 separate output voltages. Both types (or anything in between) show up in the block diagram as a single box called Power Supply.

Each labeled box in the final block diagram should be a functionally separate part of the overall project. This isn't to say that all of them are simple designs. If you need a complex power supply, for example, you might have to treat it as a separate project—that is, break it down and create a block diagram for the design of the power supply alone. Sticking this kind of detail in the block diagram for the overall project would be a mistake because it would obscure the layout of the project. When you run across something like this, the name of the game is to keep things in. Don't leave things out, but don't put extra things in.

If you're working on digital circuitry, the other sections might be master clocks, counters, divide by X, decoders, and so on. Because you are the one designing the project, you are the only one who knows what has to go into it. You should be saying things to yourself like "I need one of these because . . . " and "The only way I can make this happen is to put a . . . between this and that."

The same analysis applies for analog stuff. You'll find yourself drawing boxes with labels like Amplifier, Oscillator, and other things. Obviously, the names only have to have meaning for you; and if you feel more comfortable with long descriptions than single words, go for it. Whatever works. Hey, it's your design; and since this is America, you can have it any way you want it.

Once you have identified the individual sections and have drawn each of them on paper, you can start adding the arrows to show how they're connected. In general, two types of things are indicated by the arrows: single connections such as a signal or control line, and more complex connections like data or address buses. There's no hard-and-fast rule about how these should be drawn. A suggestion, however, is to indicate single connections with a thin line and complex ones with thick lines. If the connection is bidirectional, put an arrow on both ends; if not, don't.

All of this stuff is common sense; and because you're doing it to make your life easier during the design, you should adopt any convention you're comfortable with. Some people like to use different colors to indicate different things, and I understand that cross-hatching is really in now. I once worked with someone who was big on sewing thread and colored construction paper. He didn't last very long, but I have to admit his stuff was beautiful. Totally useless, however.

One word of caution when it comes to making block diagrams. It's really easy to get so caught up in this part of the project that you start to see it as an end in itself.

I have a friend who is a total freak about photography. He has a camera collection that takes up several rooms in his house, and he can tell you the particulars of every lens and camera ever made. What he can't show you is a bunch of photographs. He's gotten himself so hung up on the hardware that he's completely forgotten what you're supposed to do with it.

Cameras are used to produce photographs, and block diagrams are used to produce circuits. Both are only tools, and the only reason for their existence is to take you further into the project. Generating neat paperwork is a good thing, but if that's all you have to show at the end of the project—well, let's just say that you've come up a bit short.

A block diagram is a prenatal version of working circuitry. Even if you do it in oil paint and hang it in the living room, it's only worth something if you use it.

Designing, the fourth step

So this is finally where you get your hands dirty. All the brainwork you've done previously has been aimed at getting you to this point, and you should have a reasonably clear understanding of what you want to accomplish at the bench. Even though the paperwork you've been doing up to now was supposed to have been on a more abstract and theoretical level, you can't help having a few practical thoughts sneak in when you're not looking.

If you've been following all the steps leading to this point, you might think that you've only been doing brain-type stuff; but nobody can spend paper and pencil time without thinking about how it's all going to be translated into reality. Even the most cranial people have been known to have a thought or two about how they'll lay out the design, select the wire to use, get all the components together, and so on.

Even though you're working on a single project, each box in the block diagram is a separate circuit, and every circuit has to be designed and working before you can begin to refine the complete project. While it's certainly true that you have to keep a kind of global overview of the project as you're working, it's also true that you have to concentrate on each part when you're designing it. As we'll see in a bit, however, a few extra complications can develop when you're designing a piece of a larger project that don't arise if the circuit you're working on is all the circuit there is.

As we get into individual projects later in the book, we'll be going through the entire design process for each one. You'll see that while each step of the process is really a separate job, there's a lot of mixing around the edges. That's a good thing because the idea behind the whole process is to translate your first idea into working hardware. But nobody is that totally mechanical—there's a big difference between being methodical and being a robot. While you may have complete control over what you're doing, unless you're a Martian or a mutant, there's no way to control what you're thinking.

When the time comes to start designing hardware, you're bringing to the bench a lot of completed paperwork to guide you through building the circuit. The two main reasons for spending so much time with paper and pencil were to add structure to your idea and to figure out the best way to go about designing it. As you were generating the final version of your block diagram, specific hardware thoughts should have been creeping into your mind.

These thoughts are usually notions about ways to handle particular hardware problems. I'm calling these "notions" because they obviously haven't been completely thought through. (After all, you were spending your time on the overall layout of the project, not specific parts of the design.) These notions can turn out to be valuable, however, and throwing them away would definitely be a bad idea. The right thing to do is to write notions down as they occur to you and go over them when you actually start designing the hardware.

If you're the kind of person who starts a design without spending any time at all working out the details, you have to have had feelings of being overwhelmed when you sit down at the bench. The first thing that should cross your mind is where to start. By doing this planning ahead of time, you'll find it's much easier to get started because the block diagram you've created is a skeleton form of the schematic of the complete circuit. Looking at it tells you exactly what kind of sub-circuits have to be designed; and because you know what they are, you can be almost free to choose the order in which to design them.

Where to start a design depends on the complexity of the complete circuit and the interdependence of the individual sections. I always start with the power supply, but that's a choice based on my own personal superstition rather than any electronic ground rule.

The only real consideration you should have as you're starting the hardware part of your design is doing things in an order that makes it as easy as possible to test sections as they're completed. The power supply I traditionally begin with is a perfect example. No matter what your supply is supposed to do, it's a simple matter to see if it's doing it. Unless your supply is supposed to provide weird voltages that have to be locked in some complex mathematical phase relationship, it doesn't take much work to find out if it's doing what you want it to. This is true for any supply, regardless of whether it's just a pair of capacitors or a fully regulated, active circuit that generates a range of voltages with goodies like current limiting and short-circuit protection.

The big difference between designing a stand-alone circuit and one that's a section of a larger design is that stand-alone circuits are usually simple designs that don't have to work with anything else. To rephrase Gertrude Stein, a light dimmer is a light dimmer is a light dimmer.

When you design circuits intended to be integral parts of a project, there are other things to keep in mind besides making sure the thing is working in the first place. You have to be sure that the circuit can talk properly to the rest of the sections of the overall project. This can be stuff like impedance matching, line levels, signal detection, and so on—it all depends on what you're building.

The bottom line is that designing a piece of a project is not the same as designing the whole project all at once. You shouldn't move from one part of the design to another without knowing absolutely that the part you've just finished is working correctly. Likewise, there is a best order when you're working your way through a complex design. The exact order depends on the project but, in general, you should start off with the parts of the circuit that will be needed to help you test the other parts.

All this stuff is theory and, while it's usually true, that's not the same as saying that it's always true. Circuit design is as much an art as a science—and that leaves loads of room for expressions of individual preference on the part of the designer.

What we'll be doing in the rest of this book is going through the complete design of projects and seeing how the things we've been talking about so far get put into practice. We'll be designing useful circuits—things that you can stuff away in the back of your brain and hold there until a need for them arises. But no matter how handy you find these circuits, remember that how the designs were done in general is just as important as the components used. Method may not be everything, but it's up there.

The trouble with theory is that you can only learn things theoretically. It's much more useful (and a lot more fun) to put the theory into practice and actually build some real-world stuff.

So roll up your sleeves, sharpen your pencils (does anyone do that anymore?), and let's start our trip into circuitland.

2
Building blocks
Looking under the hood

A popular notion that gets drummed into people's heads is that a first impression is the best impression. I'd like to meet whoever came up with that idea. While it's not bad in theory, from a practical point of view it leaves something to be desired— a whole lot of something. First impressions are trustworthy if you're talking about stuff that's one-dimensional—hot things are hot, cold things are cold, and you always find lost things in the last place you look for them.

Think about that. The more complex something is, the more you should treat your first impression of it with suspicion. Subtlety and nuance are also often overlooked at first glance. If you had to formulate a general rule about this kind of stuff, it would be something like

If it takes time to build, it takes time to understand.

Some things are so complex that people can, and often do, devote their entire lives to understanding them.

Electronic design is as much art as it is science. Take a pair of equally knowledgeable designers, give them the same design criteria, and they're bound to produce two different circuits. Both will do the job, but each will get it done in his or her own way.

If you want to be a successful designer, give more freedom to your creative energy. If you let your mind wander through a problem, there's no way you'll come up with a workable solution. Something that works may appear on the bench, but more than likely it won't be anything workable.

You can divide the development of a circuit lots of different ways, each giving you a different perspective into just how the whole process is supposed to work. We've already thoroughly dissected the mechanical steps, but there's another approach that can only be done once you understand just how methodical you have to be when starting a project.

There's no getting around the importance of method. You've got to follow the three basic steps—from the list of criteria, through the block diagram, and finally to the hardware—to maximize your chances of success. To get through these steps, you have to approach them with an analytic eye. A coldly dispassionate, scientific view of the job is your best friend in the early days of the design.

But things change. The more clearly defined your goals become through the process of logical analysis, the more attention you have to give to the creative side of your brain. When you have a bunch of logical facts cooking in the back of your mind, they don't always combine to produce exclusively logical answers. Never forget that:

Your subconscious is your best friend.

It will suggest possibilities to you that you just can't get from a strictly logical analysis of the facts.

While you're in the middle of a project and find yourself totally ossified by a problem, if you're lucky—really lucky—you'll go to sleep with the problem kicking around in the back of your brain and wake up at three o'clock in the morning screaming, "Aha!"

It really does happen like that—really. At least as far as your work life goes, it's the greatest feeling there is.

Just after Heisenberg published his Uncertainty Principle, he was asked how he arrived at his final conclusion because there was nothing in his preliminary work that led directly to it. He said that he went to sleep with the problem stated on paper and woke up with the answer in his mind. Even though he hadn't worked out the proof yet, he knew the answer was right because he believed in "the symmetry of the universe."

Now that's pretty heavy stuff, and the problems you'll probably be working out at the bench will be slightly less cosmological (but no less important). The methods, however, are exactly the same.

Successful circuitry is as much the result of inspiration as it is perspiration. You have to dream just as much as you analyze. The secret is knowing when to do which. While every project starts out with a purely analytic approach, the further you get, the more you'll want to mix logic with imagination.

The angelic point of view

One of the biggest problems faced by designers who don't have a lot of experience is not being able to see the big picture. The deeper you get into a project, the harder it is to remember what your overall goals are. You can only concentrate on one thing at a time. When you're struggling your way through a small problem with one small part of one small box in the block diagram, it's tough to keep in mind that there's a lot more to the project.

Every successful designer I know has learned how to maintain the "angelic point of view." This is a mildly schizophrenic attitude in which you can bring your attention to bear on particular problems while another part of your mind floats above the bench, keeping an eye on your progress and constantly evaluating how what's happening fits into the overall project goals. No matter what you're working on, the objective is to satisfy the design criteria and build a working version of what's on the bench.

It takes a good deal of experience to be able to split your brain in half like this; but the more time you put in at the bench, the more natural you'll find what undeniably is an unnatural attitude. Even though you're undoubtedly aiming toward a working circuit, what you're going to wind up with is a prototype. It's important not to forget that a prototype is nothing more or less than a collection of stuff that does what you'd like your circuit to do. It's a step along the way, not the final word. That's what revision numbers are all about.

The goal you should have in mind is to make something that works, not something that's perfect. Sure there's a better way to design the amplifier, different clocks put out much squarer pulses, and now that you're working on this part of the design you realize what you can do to improve an earlier part you've already finished. Write down your ideas, but don't even think about doing anything with them.

The big difference between developing a prototype and producing a product is the difference between designing and refining. *Prototyping* is the designing of something that never existed before while *refining* is the process of making it better, faster, cheaper, smaller, and so on while keeping it functionally the same. There's a similar difference between writing and editing. Products start out as working prototypes, but prototypes start out as good ideas.

If you're an experienced designer, you can maintain the angelic point of view because it's easy to see each box in the block diagram as working hardware. You can't see what the final version of the hardware is going to be, but you know what you have to do to make something that works because you have a good set of notebooks filled with subcircuits that you've used before and know are reliable. The better your notebooks, the faster you'll be able to get the prototype working.

Because the experienced designer doesn't have to devote all of his or her concentration to creating the initial circuitry from scratch, it's a lot easier to focus on the overall project goals and make plans for improving the circuit once the design is finished.

Getting information

There's no way that a book (even this one) can substitute for years of experience on the bench, but you can get a good head start by having a bunch of handy-dandy subcircuits to draw on. There are a lot of books on the market—including big, fat ones—that are collections of circuits for everything you can think of, but all of them are short on theory and devoid of explanation.

A much better place to find basic subcircuit information is in the databooks put out by the semiconductor manufacturers. These books have basic configurations for most of the ICs and, more important, list every chip's known operating parameters so you can see whether a particular chip is really the one you want to use. Trying to design without a databook is dumb.

The goal of a semiconductor manufacturer is to make products as easy to use and as usable as possible. This goal is terrific, but it frequently causes a problem when the two objectives don't necessarily coincide. Making something as useful as you can means making it able to do lots of different things. The classic case is that of a microprocessor. I can't think of any piece of silicon that's more flexible, but I can certainly think of other silicon that's easier to use.

When you're ready to design with ICs, databooks really have all the information you'll ever need but—and it's a big but—they usually assume you know some stuff beforehand. The only pieces of writing you'll ever find that are worse than a databook are the instructions that come with tax forms. Fortunately for all of us, IC manufacturers also publish application notes. These are practical applications that go into great detail about how to use a particular IC in a particular kind of circuit. You should have both types of books in your library.

Another great source of information that people overlook when doing research on a project is the United States Government—the world's largest publisher. Believe me, you want to be on this mailing list. The U.S. Government Printing Office (GPO) publishes information on every subject you can imagine and lots that you can't. If I wanted to build something that required information I didn't have, one of the first places I would turn to is the GPO catalogs listing government publications. Next to the old catalog that used to come from Edmund Scientific, the GPO catalogs are the ultimate books to leave in the bathroom for occasional browsing.

If you're really into electronics and plan on spending lots of idle hours at the bench, you've got to make a habit of collecting as much information as you can about everything you can. You never know where you'll find something that's useful for a design somewhere down the road. Remember,

All good information is worth something.

The difference between success and failure can be one obscure fact you picked up by browsing through an issue of *Modern Dentist* the last time you went to have your teeth cleaned.

There's a corollary to this rule that you might already know. In case you've forgotten it or don't give it a lot of importance, keep in mind that

All bad information is worth nothing.

Besides being wrong, one bad fact can set you off on a path that leads straight from nowhere to noplace. By the time you realize that you've been thrown off the track, you'll have wasted a lot of time and smoked a lot of silicon. If you're designing for a living, you've wasted your client's money and stand a good chance of losing the job.

Designing electronics is a creative business. It's more art than science. Every project you complete is a reflection of who you are and what you know. The more rounded your education, the broader your interests and the more ammunition you'll have when you're working to bring an idea to life. A good designer is interested in everything there is. Your mailbox should be filled—every day—with literature, magazines, and catalogs on everything. Stay current on stuff.

The more you read, the better you'll get at spotting the difference between good and bad information. How well you can do that is the real measure of how well you're going to do at the bench. You may think that there's absolutely no connection between electronics and, say, carpentry or animal husbandry or rebuilding player pianos, but you're wrong.

Your conscious mind may find it hard to make a connection between spayed cats and infrared transmission, but your subconscious (where all your creative leaps are made) doesn't prejudge stuff. I don't have the vaguest idea how the subconscious mind works, but I've learned that the more information—good information—it has to work with, the more it can do for you. And it can do a lot.

Getting a head start

Without the use of certain Krell equipment and a few pieces of gear developed by Klant Zorch, there's no way to get 10 years of experience without going through 10 years of work. You don't start out at the bench with the ability to see things at a glance until you've glanced at lots and lots of stuff. Translating block diagrams into working hardware, while maintaining the angelic point of view, is like red wine—something that improves with the passing years.

One thing that does come with years of experience is a notebook collection that has lots of pages filled with subcircuits, tips, techniques, and, in general, all sorts of stuff that you've used in the past and found to be reliable. The notebooks contain the designs for particular circuits, but they're handy to keep around because you know you'll be using them over and over in other projects.

Every project has its own peculiarities and demands; thus a subcircuit that's perfectly tailored to meet those needs will more than likely not be perfect for anything else. But that's not to say it can't be slightly modified or even left temporarily as it is to work somewhere else.

The notebooks you develop are the most valuable resource you have, and your success as a designer is linked directly to how well you keep notes on what you do. They're the first place to look when you start working on something new and the only place to look when you're revising something you've already done.

What we'll be doing as we go through this book together is learning the right way to design—how to take an idea and make it come to life by following the formal steps that lead through the occasionally bewildering process of design. We'll learn how to start out with nothing and wind up with something.

Since notebooks are such an important part of the process, we'll get started by doing some simple designs that should be included in your notebooks. If you don't have one, the circuits we'll develop are a great beginning; and if you already have one, these circuits will be a great addition.

Subcircuits

Let's start off by defining our terms. A *subcircuit* is a circuit that does a particular job needed in a larger circuit. A clock, decoder, converter, or amplifier is an example of a subcircuit. When you draw a block diagram of the project you're working on, each box in the diagram is a subcircuit. Subcircuits do jobs necessary for the functioning of the complete design, but individually they don't do things that are really useful as stand-alone projects. If this is a bit confusing, you'll get a feeling for it as we go through the work of designing a few of them.

To a large extent, the difference between a subcircuit and a complete circuit depends on your point of view. If you're building something that has to control the lights and heating in your house, you'll need an accurate clock, relay controls, and some light and temperature sensors. Each of these is a subcircuit; working together, the subcircuits make your home controller.

But if you were bent on designing an accurate clock for the wall, the same clock that was a subcircuit for the home controller would be the complete circuit for the wall clock. What you call something depends totally on what your final goals are.

But enough talk, let's do some work.

Operational amplifier power supplies

The operational amplifier, or op amp, supply is a classic example of a subcircuit. I'm referring to this subcircuit as an op-amp thing because the op amp is usually where it shows up. However, like any subcircuit, it can be used in other designs as well.

Op amps are basic circuit building blocks, but one of the frequent hassles that you have to deal with when you use some of these chips in a design is their power requirements. Most of the common op amps use a bipolar supply. This means that you need a voltage that's really negative with respect to the system ground.

You can fool an op amp by using capacitors and/or resistors to produce a negative reference, but these are phony-baloney techniques that carry so many operating restrictions you're better off forgetting about them entirely. It's a lot better to look for an op amp (and they're out there) that is happy with a single-sided, positive-only power supply.

But that's not the way to approach design because limiting the choice of components means you can't make the circuit work exactly the way you want it to. If an op amp (or something else) wants a true negative voltage, the best thing is to give it what it wants. An op amp like the 741, for example, can be made to operate with a single-sided supply, but its performance will be degraded. This may not be such a big deal in your application, but being willing to accept second-rate performance is a lousy attitude to have and will probably affect the performance of the entire project. It's not a good idea to accept second-rate stuff, especially when it's not necessary.

Generating a real negative voltage is easy. It would take a bit of brain strain to be able to produce mega-amps of current but, as is the case with op amps, if you're only looking for a handful of milliamps, the answer is simple.

The basic technique behind getting something below system ground level is shown in Fig. 2-1. This design has been around for years and can be used for lots of things besides the care and feeding of op amps and other chips that need a negative reference voltage. The method shown in the schematic is called a *charge pump*. Once you get familiar with it, you'll find it to be a real lifesaver (or as much of a lifesaver as any subcircuit can be).

The charge pump has to be fed with a clock, and square waves work just as well as anything else. On the positive swing of the input clock, D1 is forward biased and D2 is reverse biased. This causes C1 to charge up and the amount of charge depends on the capacitor value, the voltage swing of the clock, and the output frequency.

When the input clock moves on to its negative half-cycle, C1 dumps its charge through D1 and charges up C2. Since C2 is receiving its charge through its back end (so to speak), it can't discharge through D2 because the potential is lower at the C2/D2 junction than at the C1/D2 junction. The process repeats on the next clock cycle.

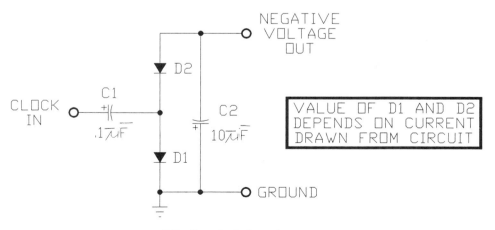

2-1   Standard charge pump.

The result of transferring the charge from C1 to C2 (through the two steering diodes) is that a voltage shows up at the C2/D2 junction that is negative with respect to system ground. Before you run out and try this circuit, there are a few things to keep in mind.

There's going to be a lot of noise at the output of the charge pump in the form of residual ripple due to the constant charge/discharge action of the capacitor. You can filter and regulate the voltage just as you do with any supply; and the higher the frequency of the input clock, the smoother the raw output voltage is going to be.

There's a limit to the frequency, however, because too short a cycle won't let the capacitor charge completely and this will lower the amount of current you'll be able to get from the pump. Too low a frequency will result in an output waveform that's so choppy you'll really be looking at a clock instead of a constant voltage.

There's no easy answer to the frequency you should use to feed the charge pump because it depends on the load impedance of whatever the charge pump is powering. If you pull a lot of current from the circuit, you should look more at bigger capacitors and less at a higher frequency.

The values shown in the schematic will give you an output voltage of about 70% of the maximum voltage swing of the clock you use to drive the circuit.

Although you can use any clock to feed a charge pump, you want one based on something that can supply a fairly meaty amount of current because the more you can put in the front end, the more you can take out the back. An ideal choice for the clock is a 555 because it can supply about 200 mA at its output, is easy to use, and produces a clock with output swings that almost reach the supply rail. Figure

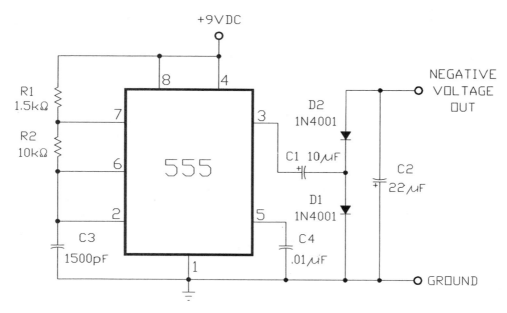

2-2   A 555-based negative voltage supply.

2-2 shows a working version of a negative supply circuit. For most of the uses you'll have, you'll probably find the characteristics of the circuit to be adequate.

If you power the circuit with a 9-V battery, you'll be able to get about – 6.5 V at the pump's output with a maximum current capability of somewhere around 30 mA. The 555 is configured to produce a frequency of about 500 kHz, which results in minimal residual ripple noise. You can get more current out of the circuit by raising the capacitor values and lowering the clock but, as we discussed earlier, there are limits.

Remember that the circuit will work with just about any input clock. If all you need a negative voltage for is a simple reference for an op amp, there are other clock sources you can use. Since an op amp only needs flea power on the negative side, you could build a simple oscillator from extra inverters or (and this is really slick) you can steal any handy clock that you already have operating for some other purpose in the circuit.

As with all the subcircuits shown here, you should build a charge pump and experiment with it. See what kind of output you get with various input clocks and what the relationship is between output noise and capacitor values.

If you really push this technique to the wall and start looking for ways to get serious amounts of current, you'll have to use heftier diodes. A word of caution— the bigger the capacitors you use, the larger the charge you'll have. When you start getting into three-digit amounts of microfarads (and even the upper two-digit values), you're starting to generate charges that can do a whole lot of damage if they discharge by accident. Circuit components can be replaced, but fingers and stuff can take a long time to heal.

Frequency doubling

Just about the most common circuit activity you'll ever run across is frequency division. Modern circuitry spends a lot of its time taking a master clock (of some enormous frequency) and running it through a series of counters and other chips to slice off frequencies whose relationship to the master clock is a matter of simple arithmetic.

Most modern radios, televisions, and computers use this technique to generate the frequencies they need. If you spend some time with the schematic, you'll see that there are lots and lots of counters all over the boards.

A less common activity, but one that occasionally is necessary, is frequency multiplication. Although there are circuit techniques to do this (and we'll look at a really snazzy one later on), getting it done is not a trivial thing—at least not the sort of thing that you can toss off between breakfast and lunch.

Everybody knows how to design a circuit that will take a frequency and divide it in half. This is such a basic piece of business that a lot of schools use it in class as an introductory exercise in design techniques. What isn't mentioned—probably because there aren't any handy-dandy chips around to do the job—is how to take a frequency and multiply it by 2. This activity isn't as widespread as frequency

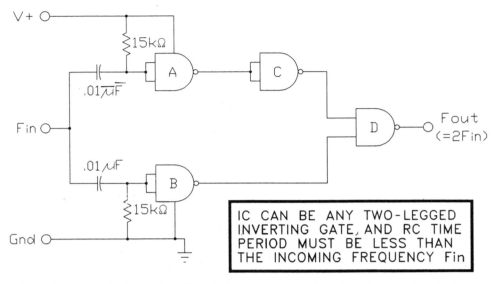

2-3 A frequency doubler. The RC period must always be less than the period of the incoming clock you want to double.

division, but it comes up often enough to have a simple subcircuit worked out in your notebook and available for use whenever you need it.

There aren't any common chips designed specifically for frequency doubling (and why that's true is a good question), but you can take some common chips and set them up to do the job.

The subcircuit in Fig. 2-3 may well turn out to be the most useful one in your notebook. Ever since I designed it a couple of years ago, I've found myself dropping it into the middle of other projects. It has saved lots of time, and just about the only restriction I've ever had to overcome was the rate of the clock. I've shown it built around a 4011 NAND gate, but you can use any two-legged inverting gate.

The frequency limitation of the circuit is really just the maximum operating frequency of the IC. I've used a complementary metal-oxide semiconductor (CMOS) IC in the schematic and, as with any CMOS IC, the maximum frequency the circuit will handle is dependent on the supply voltage. The relationship is pretty much linear in that the maximum frequency will double if the supply voltage doubles.

This used to be a big problem with CMOS stuff because it couldn't match transistor-to-transistor logic (TTL) for operating speed. The introduction of high-speed CMOS—the 74HC and 74HCT families—has finally brought modern CMOS parts up to speed—so to speak. Sorry about that.

The operation of the circuit is simple. Gates A and B have been configured as edge detectors so that A detects the negative transition of the input waveform and B detects the positive transition. These two gates are set up as half monostables. We'll talk more about them later; for the moment, let's just say that the inverters square up the RC pulse generated by each of the resistor/capacitor pairs at the input of the circuit.

The output of gate A is inverted and sent to one leg of the remaining gate in the package. The other leg is fed directly by the output of the gate B edge detector. If you work out the truth table of the circuit, you'll see that the final gate changes state twice as fast as the input clock because it reacts to the pulses generated by each edge detector at the input. In other words, the output frequency is exactly twice the input frequency.

When you use this circuit, keep in mind that the inverting chip has to be able to operate at twice the frequency you're going to be multiplying. Schmitt trigger parts such as the 4093 quad NAND gate will work a bit better because their built-in hysteresis tends to make the circuit more immune to noise.

NiCad batteries

Most of the projects I get involved in designing have one criterion in common: they have to be able to be powered by batteries. In more than half these cases, it's also specified that NiCad batteries have to be used. This is usually because the final design is going to be sealed inside a box, and the manufacturer doesn't want the end user to have to change batteries all the time.

Until something better comes along, that means NiCads. Sealed lead gel cells are a good alternative, but I've only occasionally been able to talk the people paying the bills into switching over to them. For what it's worth, I like them better than NiCads and use them in most of the battery-powered stuff I build for myself. They're a lot more forgiving about sitting on the shelf, have high energy density, are inherently self-regulating, and don't have the memory problems you find with NiCads. The down side is that, for some reason I don't understand, they're murder to find in onesies.

In any event, all rechargeable batteries have to be charged and, as you know, they always run out of energy at the worst possible and absolutely most inconvenient time. Some years ago I got tired of this happening and decided to design something that would monitor the state of the battery as well as regulate the charge rate to cut the charging current as the battery absorbed more and more energy.

The only three things I had on my list of criteria were that the circuit had to be really small, use the smallest amount of power it possibly could, and illuminate light-emitting diodes (LEDs) to let me know what was going on with the batteries.

After throwing away more designs than you can imagine—mostly because of the amount of power they consumed—I came up with the idea that I would only light the LED when the battery was low. Every previous design I had come up with worked the other way—the LED stayed lit as long as the battery was charged and went out when it was so near death it had to be recharged. As soon as I made that basic design decision, everything fell into place and I was able to get working

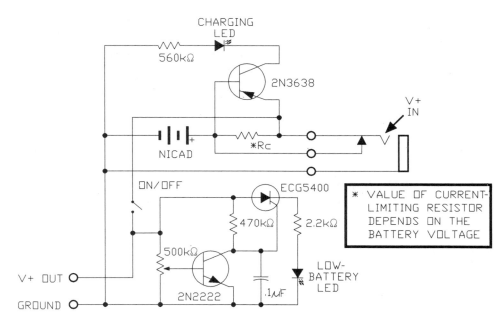

2-4 A charge and voltage monitor for NiCad batteries.

circuits that drew currents down in the microamp range. The schematic in Fig. 2-4 shows the result of this midnight realization.

The circuit has two basic parts. The first monitors and controls the charging of the batteries, and the second monitors the amount of charge left in the batteries when they're being used to power the circuit.

When you're charging a NiCad battery, the most important choice in the circuit is the value of the current-limiting resistor. Too large a value will result in no charging action, and too small a value will allow so much current to run through the batteries that they'll either be damaged or, if you're not lucky, destroyed. All NiCads produce and absorb oxygen during a charging operation. If the charging rate is too high, the gas will be produced faster than it can be absorbed, pressure will build, and the battery will explode. Really explode.

In the course of developing this circuit, I tried an early version on a NiCad and didn't notice that I had much too low a value for the current-limiting resistor. Ever notice how little difference there is between the color coding for a 1-kΩ and a 12-Ω resistor? A minute or so after I powered up the circuit I went to get a cup of coffee and, on the way back, I heard a small explosion—about the same as you'd hear from a cap pistol. (Are they still around? the pistols and the red rolls of caps?) The circuit had sent so much current through the battery that the force of the gas pressure had blown one end off the battery. The end had been moving so fast it had hit my soldering iron and knocked it over. Be warned.

The pnp transistor has its base-emitter junction sitting across the current limiter and, when current flows through it, a voltage drop appears across the resistor.

This causes current to flow through the collector-emitter junction of the transistor, and it lights the LED to indicate that the battery is charging. This isn't just a pilot light for the charger. If current isn't flowing into the battery, you won't get a voltage drop across the resistor; and the LED, as you would expect, won't light up.

As the battery voltage increases, the battery impedance increases and the current flow gets less and less. This results in a constantly lowering voltage across the resistor; in turn, the LED gets dimmer and dimmer until it finally goes out completely. The LED, therefore, is giving you a visual indication of how well charged the battery is. If the intensity of the LED doesn't change, you're getting some good information as well. Constant high intensity means the batteries aren't taking a charge, and a low intensity when you first plug in the charger means the batteries were already charged.

The second part of the circuit monitors the charge on the battery while the batteries are being used. The full battery voltage is put across the potentiometer and a certain value appears at the potentiometer's wiper. The npn is set up as a switch; and as long as the voltage at its base is high enough to keep it turned on, the collector-base junction conducts and keeps the collector at close to ground level.

When the battery voltage falls below a level determined by the potentiometer setting, the transistor turns off and the battery voltage appears at the collector. This triggers the silicon-controlled rectifier (SCR) and lights the LED to warn you that the battery has to be recharged. The SCR latches and the LED stays lit until you turn off the power.

The critical element in the monitor circuit is the setting of the potentiometer; thus, you should use a multiturn potentiometer. The circuit can be calibrated by hooking it up to a variable power supply and setting the voltage to the value you want for the trigger voltage. With NiCads, this is one volt per cell—two-tenths of a volt below their nominal voltage when fully charged. A 6-V NiCad pack should be considered to be in need of a charge when its voltage drops to 5 V. It's a matter of simple arithmetic.

The schematic shows the on/off switch for the circuit being powered and uses the single-pole, single-throw switch in the jack to change the power source from the batteries to the charger. If you trace through the connections, you'll see that the charger recharges the batteries and powers the circuit at the same time. Make sure you use a power source that can supply enough current for both. Once again, it's a matter of arithmetic.

The final version of the subcircuit I came up with is one I've been using in every battery-powered device I design. It's never caused any problems and is one of the most useful subcircuits you can have in a notebook.

Pulse conditioning

Considering the state of modern electronics, you can build just about anything you can think about building. The only real limitation is your own imagination; skill and experience are always guaranteed to come with time, but ideas are something else. Now I'll admit that stuff like warp drive, nitron beams, and the reverso ray designed by Klant Zorch on Neptune may be difficult, but that's more a matter of getting the parts. The heart of the device is an alloy of unobtainium.

No matter what you build, you're always going to have things that produce trigger pulses and other things that respond to those trigger pulses. Half the battle in getting something to work is making sure that the pulses being generated can be detected by the circuitry that you've designed to do the detecting.

The phrase that's thrown around for this is *pulse conditioning*. A good deal of your time at the bench is going to be spent on the design of circuits that get pulses in shape (sorry for that) for some other circuitry further down the line. This includes lengthening or shortening pulse widths, increasing or decreasing voltage swings, and other fun activities.

You can save a lot of bench time by having a handy collection of reliable circuits that will take a pulse in at one end, change it in some way, and then spit it out the back end. You should be able to put this type of subcircuit together without thinking twice—hopefully without even thinking once.

There's no end to stuff you can do to pulses—from making them get really narrow (called a *spike*) to extending them out so they'll last for hours (called tedious). The most common things done to pulses are to stretch them out and delay them. These two activities are at the bottom of most pulse-conditioning circuits, and they're good ones for you to have in your back pocket when you sit down at the bench to work on a project.

Pulse stretchers are used for a variety of things in most of the circuits you'll ever build. On an abstract level, all they really do is detect the incoming pulse and then generate a wider pulse of the same polarity. There are several ways to do this and, to a certain extent, the way you go about it will depend on the characteristics of the pulse you're detecting as well as the type of pulse you want to generate in its place.

It's always a good idea to keep these circuits as simple as possible because the output pulse is always delayed by a certain amount due to the propagation delays of the components and whatever RC time constants are included in the circuit. Another good reason for keeping things simple is that the fewer parts you need, the more chance you'll have to build one from some unused silicon you already have on the board.

Figure 2-5 shows a versatile pulse stretcher. As you can see, the pulse stretcher has two sections. The input section detects the incoming pulse, and the output section produces a new pulse. The circuit is a really simple one. The input pulse detector is nothing more than a simple inverter (gate A) that sees the incoming pulse and outputs an inverted version of it.

The inverter can actually do more than that if you choose your parts carefully. A simple inverter will take a weak pulse with a minimal voltage swing and output a nice healthy pulse. How healthy the pulse is depends on which gate you use and which logic family it belongs to. If you use a Schmitt trigger version of an inverter, like a 4584 (highly recommended), the gate will also clean up noisy pulses with jittery rise times.

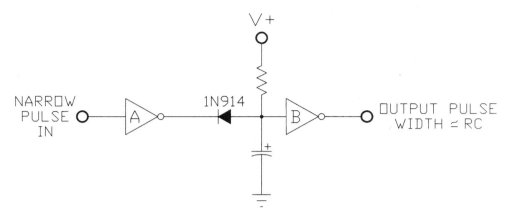

2-5 A pulse stretcher for positive-going pulses.

If you're detecting a positive pulse, a negative pulse will appear at the output of the inverter. This pulse will discharge the capacitor through the diode, and the output of the second inverter will go high. Even if the input pulse is very short and causes gate A to change state before the output pulse has been completely generated, the diode isolates the capacitor from the positive output of gate A and keeps the capacitor from charging up.

You'll notice that this pulse stretcher only works with positive pulses. Doing the same thing for negative ones is easy. You could put another inverter at the input, but this would reverse the polarity of the input and output pulses. Figure 2-6 presents a better alternative. Notice that the only difference is that I've exchanged the positions of the resistor and capacitor and have turned the diode around.

The width of the output pulse generated by the circuit is roughly equal to the simple RC time constant. Actually, it's a bit less due to the presence of the output inverter, but the goal isn't super-duper accuracy. If you're looking to produce

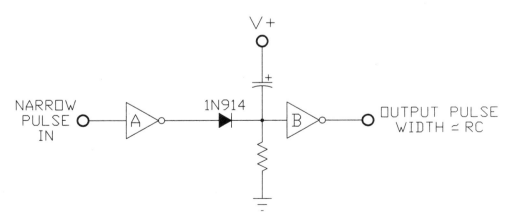

2-6 A pulse stretcher for negative-going pulses.

output pulses of a strictly predetermined width, you'll need a much more sophisticated circuit with parameters that depend on the rest of the project you're designing.

You can use 555s to build a terrific pulse generator (monostabl. multivibrator);
and if you have something like half a 556 unused in the project, it's a great way to
go. Just feed the input pulse to the trigger input and that's about it. The trigger
wants a negative pulse and the output will be a positive one, but you can use inverters to take care of that. You can also trigger the 555 with a positive pulse if you
hold the trigger input (pin 2) low and use the reset input at pin 7 for the trigger.
Remember that the 555 operates when the reset input is held high. The disadvantage of doing this is that the output pulse will end abruptly if the reset input goes
low before the 555 has timed out. I've drawn both of these conditions in Figs. 2-7
and 2-8. If you know that your circuit won't reset the 555 before the output pulse
has completed, it's a reasonable way to go about stretching pulses.

The inverter circuits are faster, more versatile, and can usually be made out of
spare board silicon. There are other ways to stretch a pulse, but they all work on
the same principle. If you make yourself familiar with the methods shown here,
you'll be able to design any pulse stretcher you need to meet the requirements of
whatever project you're working on.

Pulse delay circuits are easier to understand because the technique is simpler.
You can generate extremely accurate delays by using the circuit shown in Fig. 2-9.
The 4017 is an easy-to-use ten-stage counter that we'll examine in greater detail
later in the book. In its most basic configuration, you tie the reset and enable pins
low, put a frequency at the clock input, and the chip will put a high, one after the
other, on each of its outputs.

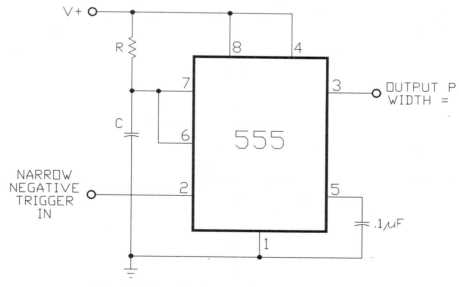

2-7   A 555-based pulse stretcher for negative-going pulses.

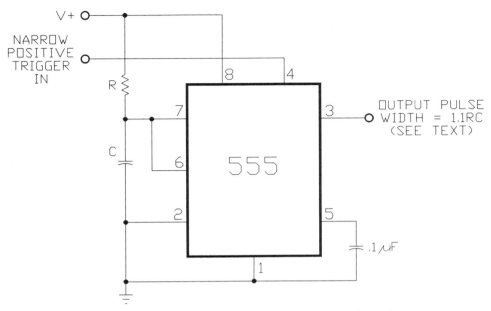

2-8   A 555-based pulse stretcher for positive-going pulses.

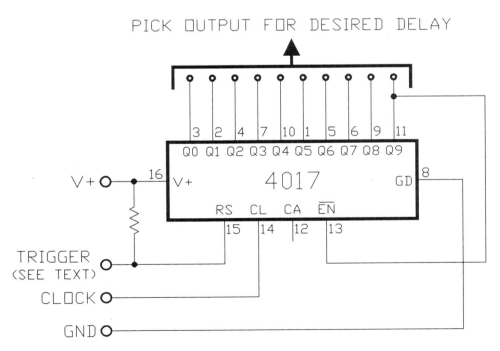

2-9   Accurately delaying pulses with a 4017.

To use this chip to generate time delays, connect the enable pin to the last output, put a convenient clock of the right frequency at the clock input, and feed the input trigger pulses to the reset pin. When the power is first turned on, the 4017 will start counting until it reaches a count of 10. When that happens, pin 11 will go high and the chip will stop counting because that output is connected directly to the enable input of the chip.

Everything will stay like that until a negative-going trigger pulse comes along, causing the chip to reset. The count starts again and continues until the last output goes high. When that happens, the chip freezes and sits there waiting for the next trigger input on the reset pin.

By being careful about the frequency you put on the 4017's clock input, you can generate accurate delays by picking your output pulse off any of the chip's outputs. This circuit is designed to work with negative-going pulses, but positive pulses can be easily handled by putting an inverter in front of the reset pin. You also have to make sure that the trigger pulse stays negative long enough for the appropriate delay to be generated. If the trigger pulses you have in your circuit are really narrow spikes, you'll have to put something in front of the reset pin to stretch the trigger pulse out. But wait a minute, that sounds familiar.

Switch debouncing

The last freebie I'm going to give you for your notebook is a bunch of ways to debounce switches. This problem, more than any other, causes beginning circuit designers loads and loads of problems because the symptoms of a noisy switch can look like lots of other, more serious, circuit problems.

Even experienced designers screw this up. I've seen many so-called "final schematics" that had to undergo a rather hasty and always embarrassing product recall because the mechanical switches and keyboards were just too noisy and the input circuitry they fed didn't properly debounce them.

There are two basic ways to debounce a mechanical switch—three if you can find a switch replacement that's quiet enough to work in your circuit. If you're debouncing momentary keyboard switches, chances are you'll have to take care of things electronically. Every one of these switches I've seen is unbelievably noisy and has to be debounced.

The two electronic methods for switch debouncing are to either soak up the extra pulses coming from the switch or, as we did with pulse stretchers, detect the first edge of the first switch pulse and use that to generate a clean pulse lasting longer than the worst case of bouncing you expect from the switch itself. Which method you use also depends on the kind of switch you're using. Toggle switches can be handled one way and momentary switches another.

Soaking up the mechanically produced multiple pulses from a noisy switch involves having a big capacitor sit at the switch output and take longer to fully charge than the longest time it takes the switch to settle down. You can do this with a simple resistor/capacitor combination, but it's smart to add an inverting or

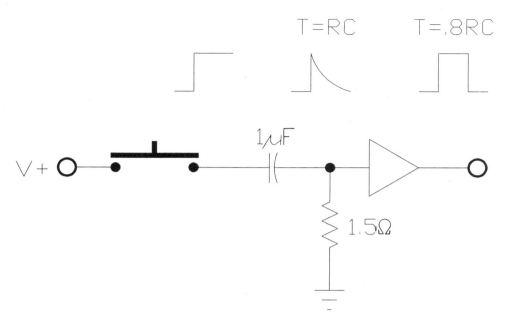

2-10   Simple RC pulses are squared up by running them through any basic logic gate.

noninverting gate to the circuit as well. These circuits are half monostables (more about them later on), and adding the gate as shown in Fig. 2-10 cleans up the very sloppy shoulders typical of RC circuits.

The length of the output pulse is equal to the RC value if you don't add the gate and slightly less if you do. The difference comes about because the gate will change state before the capacitor fully charges or discharges. Because most debouncing doesn't require exact timing precision for the output pulses, you might as well use the RC value for both circuits.

The polarity of the input pulse can be retained or not depending on whether you use an inverting or noninverting gate as shown in the two schematics of Fig. 2-11. If you flip the resistor and capacitor connections to power and ground, the circuits will respond to input pulses of the opposite polarity (Fig. 2-12). The length of the output pulse depends on what you're doing with it, but most keyboards will be fully debounced with output pulses in the range of 100 ms. This is what you'll get if you use the values in the schematics.

If you want to debounce a toggle switch—something like either a single- or double-throw switch—you can use the circuits shown in Fig. 2-13. Throwing the switch one way or the other will change the state of the output, and the resistor is there to act as a short-term storage device for the time that the switch is being thrown. No matter what kind of toggle switch you use, there's always some time during which no connection is being made. If you were to leave the resistor out of the circuit, the gate inputs would be able to float for a period of time. Always a no-no.

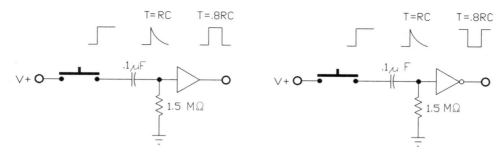

2-11 Half monostables that preserve the polarity of the input trigger pulse.

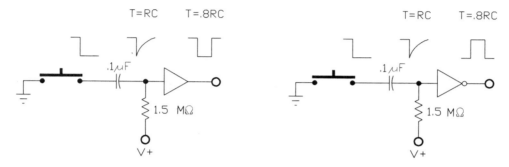

2-12 Half monostables that invert the polarity of the input trigger pulse.

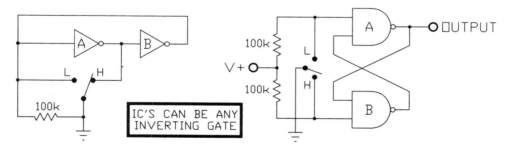

2-13 Two ways to debounce toggle switches.

The circuits shown here, or some variation based on them, will be able to deglitch even the junky switches you can get cheap from Zorch Electronics on Neptune. Write for the Zorch catalog; it's one of the most interesting ones around. You can't read the writing, but the pictures are great.

A final word

The usual method for getting stuff into a notebook is to go through the headache of developing the design and writing down all the things to watch for when the circuit is used. In general, anything that gets into your notebook should be something that you've screwed around with so much that you know every possible hassle and glitch the circuit can produce.

Getting stuff for nothing (as in this chapter) is dangerous because you only know what the circuits can do, but not all those other nasty things they can do if you don't keep an eye on them. Since the real value of stuff in your notebook is that you can use it without any aggravation, you should take the time to build each circuit and do everything you can to try and make it blow up.

Subcircuits are only valuable if you know their limits, and the only way to find out the limits is to experiment with the circuits until you can predict what they're going to do in a variety of applications. The parts are cheap, and the small amount of time it's going to take to try these circuits out is nothing compared to the brain damage you can get by using one of them to do something it wasn't designed for.

As we go through the rest of the book together, you're going to see all of these circuits (or variations) used over and over in larger circuits. I'll always explain things as we go along but the more you bring to the discussion, the more you'll be able to get out of it.

Try this stuff, and when you're sure you know exactly how the circuits are going to behave, put your brain in hyperdrive and we'll get started with some more significant designs.

A slick trick

Everybody has an occasional need for a stable oscillator, and the ultimate in stable clocks is one built with a crystal. There are several ways to do this but none is as simple as this: a couple of spare gates, a couple of miscellaneous components, and you've got yourself a good crystal oscillator.

The circuit should be built with CMOS parts, and the lead lengths should be kept to a minimum. Although I've used a 4049 for the inverter, you can use any handy inverting gates you have on the board. The higher the gain of the inverter, the higher the crystal frequency you can stick in the circuit. I've used this circuit with crystals as high as 14.318 18 MHz.

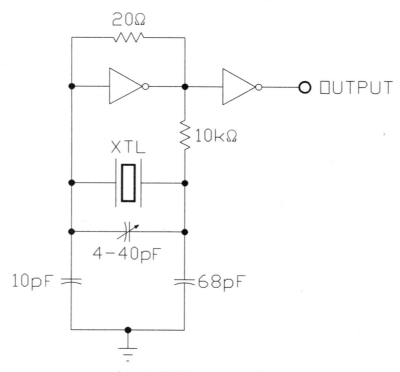

A basic CMOS crystal oscillator.

If you use a two-legged inverting gate, you can turn the circuit into a gated oscillator by having the second leg work as the enabling control. The circuit will start up without any trouble, but you should experiment with it to be sure before you add it to your personal notebook.

3
Keyboard concepts
A first design exercise

One of the unique things about electronics is what you could refer to as the "trickle-down" effect. As the years go by, more and more stuff is being built around, modified to include, or heavily advertised as controlled by—you guessed it—electronics. It may be hard to believe, but right this minute I am holding in my hand an ad for a kitchen blender that proudly carries the words (in close to End of the World type) "Solid State."

I have to tell you that I was so mystified by this that I called the manufacturer and asked about it. After playing a lot of telephone tag from one extension to another, I finally got connected to someone in the service department who agreed, after warning me that there weren't any "customer-serviceable parts" (now there's a familiar phrase) in the blender, to send me a repair manual.

To make a long story short, the new model of the blender was referred to as being solid state because there was a thermal fuse on the line. That has to be the reason because it was the only electronic component I found inside the blender.

The only reason I'm talking about this is that, as the general state of electronics moves forward, bits and pieces of the technology slip over the edge and become available to people like us. This is what I mean by the trickle-down effect, and all of us who are interested in electronics can benefit from it. Now don't think for a minute that you can call your favorite supplier and get your hands on the latest gee-whiz NASA and Star Wars stuff. If you're planning on building a 10-MW X-ray laser, I'm sorry to tell you that you're going to have to do it without the help of the friendly folks who work for your government. For some things, you're on your own—but more on that later.

There's no shortage of electronic stuff you can design and build on your own bench, and there's no reason why you can't use the most modern and recently available parts. The only problem is that all the energy in the electronics industry seems aimed at improving the general state of the technology—shoehorning more components in ICs, upping their speeds, and so on. This is really great for the end user because more powerful stuff can be made smaller and cheaper, but it doesn't really do a lot for people who like to do weekend tinkering on the workbench.

Having a large company with lots of bucks to spend on R&D develop an application-specific integrated circuit (ASIC) that contains every bit of the circuitry needed to build a TV set (except, I guess, the TV tube itself) may be a great thing for the consumer market, but there's no way it's going to do us any good.

First, we'll never be able to get our hands on any of these parts; and even if we could, the onesie price would be more than the cost of the TV set. Second, the mechanics of prototyping and designing around it would lead to an all-expense-paid vacation on the funny farm. Finally, it's probably true that after an awful lot of brain damage, all you'd have to show for your work would be a poor imitation of the TV set you could have bought in the store.

So where does that leave us? Actually, we're not in such bad shape at all. A lot of the Buck Rogers circuitry you hear about started as a large collection of ordinary components. Even the deadly Death Ray made by Klant Zorch, the renowned Chief Scientist of the fabled but long-gone fish-faced people of Neptune, was designed on protoboards with a bunch of off-the-(Neptunian)-shelf components.

You won't find one government official to confirm this, but I have it on good sources that the reason for the *Voyager* series was to try and find the crucial missing page of the Death Ray's schematics. Even the Defense Department won't admit that was really the primary goal of the recent *Voyager* mission. Try and come up with any other logical reason for taking a trip to Neptune. I mean, just what do you think caused all those planetary rings in the first place?

Extremely sophisticated, powerful, and miniaturized circuitry gets developed by starting out with acres and acres of ordinary silicon and the same passive components you buy blister packed in your own local electronics store. The biggest hassle to developing circuits on the bench is not, as you would first think, the availability of parts but rather the whole design environment.

This impressive mouthful of words refers to the test equipment you have available and the way you go about constructing the circuit. The more sophisticated your project, the more sophisticated the collection of test equipment you'll need. And the faster your circuit operates, the fewer options you'll have about how to prototype it.

The handiest, easiest-to-use, development method in the known universe is the protoboard system. Protoboards are made relatively cheap by a variety of companies and are available from all of the ever-dwindling suppliers who service the hobbyist market. You can get protoboards in several sizes and bus configurations. Two of the most common are shown in Fig. 3-1.

As we all know, however, there are a number of drawbacks and limitations to using protoboards. Some of these limitations are no big deal, but others are real stone-wall things that can make it absolutely impossible to use the protoboards for certain designs.

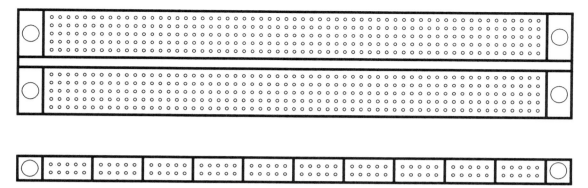

3-1 The two basic solderless breadboards used for building all the circuits in this book.

Probably the biggest problem with protoboards is that there's a real hassle with parasitic noises in adjacent buses. Take a look at Fig. 3-2, for example. These waveforms were captured on a digital scope. The top row shows a series of pulses being propagated at a frequency of 1 MHz with an amplitude of approximately 5 V. The waveforms may look a little ragged, but remember that these are real-world, actual, honest-to-gosh, waveforms—not databook theoretical drawings.

The waveforms at the bottom of Fig. 3-2 were captured at the same time by putting the probe for the second channel of the scope on the row of pins immediately adjacent to the row in which the real waveforms were captured. What should strike you right away is that each real pulse shown on the top has a phantom counterpart on the bottom. You'll also notice that the phantom pulses aren't exactly insignificant either. A quick-and-dirty guesstimate is that the phantom pulses are about 25% the height of the real ones. This is not an insignificant amount.

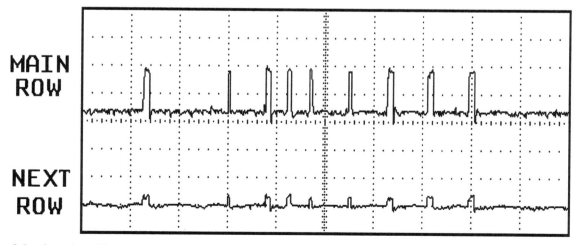

3-2 Actual oscilloscope traces showing the extent of parasitic noise in adjacent rows on a solderless breadboard.

If you're trying to design circuits that have fairly tricky timing requirements and sensitive trigger inputs, these inductively generated phony-baloney pulses can pose a big problem. You can minimize their effects by using small decoupling capacitors, but there are risks associated with this method because the same capacitors that help swallow the spurious signals have an effect on the real ones as well. The best way to keep the problem to a minimum is to avoid using adjacent rows on the protoboard.

In general, if your circuit operates below 1 MHz and you keep handy a good supply of 0.01-μF capacitors, you won't have any problems with noise.

The only other thing to keep in mind when using protoboards is to follow the same type of good construction practice you'd use with any design job. This is all really common-sense stuff such as keeping leads as short as possible, using bypass capacitors on the power buses, and so on.

All the projects we'll be doing together in this book can be built on protoboard using 24-gauge hookup wire. If there are any special considerations to keep in mind or construction details unique to the particular circuit being put together, I'll be sure and let you know as we go along. I'll also be listing the minimum requirements for test equipment and, if necessary, including scope pictures.

So let's get on with it.

Design criteria—laying down the rules

Although there's obviously nothing stopping us from designing anything we want, let's not forget that whatever we do is going to involve a good deal of time, energy, and (hopefully) brain stretching on both our parts. There's a lot to learn from any exercise in circuit design, but the creative juices flow a bit better if we know that the end product of the exercise is going to be something useful. Education should be fueled by motivation.

We've talked a lot about what you should have on a test bench. If you remember, I told you that while you can go out and buy a variety of terrific Buck Rogers stuff, there are still several things you're better off making yourself. One of these is not only a valuable addition to anybody's collection of test gear, but it's also a great first project. I'm talking about keyboards.

Now that computers have become as common as calculators, VCRs, and other high-tech consumer stuff, just about everyone past the prenatal stage of life has become accustomed to pressing buttons to get things done. Some of my friends think this activity is becoming hardwired into the genes—the next step in man's evolution. Come to think of it, people's index fingers have been getting more and more pointed lately.

A keyboard data entry system is a good thing to have on the bench. The basic design is one that can be used (in one form or other) in a lot of other circuits you may design later. Considering the state of the computer market and the fact that PC-compatible keyboards can be bought for under 50 bucks, you may well wonder why it's worth your time to design your own.

First, not all keyboards are the same. They may do the same job, but there are as many different keyboard designs as there are things that use keyboards. Computer keyboards are different from calculator keyboards, telephone keypads are different from microwave oven keypads, and so on. Each is designed to do a different job. If you take the keypads apart, you'll see that the circuitry driving each is different.

The second, and most important, reason for going through an exercise in keyboard design is that once you've done it, you'll know exactly how a keyboard works. Not only that, but the techniques used in the design should find their way into your notebook (remember that?) so you'll have time available for other designs. Now I'll admit that there's not much in the way of gee-whiz stuff in the design of a basic keyboard—and building your own is kind of like reinventing the wheel—but even if that's true, don't forget that reinventing the wheel means you'll know exactly how the wheel works. When you're into designing electronics, no matter what you're doing,

You gotta know the fundamentals.

They're part of everything else you'll ever do.

As we go through the design, you should pay as much attention to the way the design is done as you do to the design itself. The method is just as important as the circuit details. There may be a whole lot of ways to design one keyboard in particular, but there's only one way to design any circuit in general.

The first thing to do in this design—or any design you'll ever do in the future—is to have a clear understanding of what you want to accomplish. This means drawing up a list of design criteria and using that list to outline the proposed circuit in the form of a block diagram. These two steps are the first ones that have to be taken no matter what kind of project you have in mind. Remember,

It ain't what you do, it's the way that you do it.

The items that get included in a list of design criteria depend, as you can imagine, on what you want your circuit to do. This is always a personal thing because working at the test bench is usually a solitary activity. The list we'll be drawing up is straight out of my own head, but I'm going to keep it as general as possible. If you want to add anything or change any of the items on the list—hey, be my guest. All I can say is that it might be better to go along with me on the design as it is and make your changes when we're finished and you know how everything works.

If we're going through a design exercise together, it's a good idea if we're designing the same thing. When we're finished with the final layout of the keyboard, I'll give you a few suggestions on how to modify the circuit to make it more flexible.

Whenever you sit down to design a keyboard, regardless of how you plan on using it, certain criteria always wind up on the list. Let's take them one at a time.

1. The keyboard will be able to handle up to 10 digits.

Designing a keyboard that can only enter one digit is really a waste of time. Just about any application you can think of is going to need two or more pieces of data from a keyboard. Because the design of this keyboard is being done without any specific purpose in mind, it really doesn't make any sense to put an absolute, outside limit on the number of digits.

Most keyboard data systems need fewer than the 10 digits we've specified for this one; but the way things usually go, I'm sure that one of the applications you've got in mind will be needing a keyboard that can handle even more than the 10 digits I've indicated. Once you understand the way this particular circuit is laid out, and how keyboards in general are designed, you won't have any trouble expanding it to any number you want.

2. All keyboard data will be put on a latched, tristated data bus.

This may seem like a peculiar item to include in the list because there isn't much point to creating data if you don't provide a way to access it. Any data you enter from the keyboard has to exist, for a while anyway, somewhere in the circuit, but that's a long way from having a real data bus.

By specifying a data bus, and a tristated one at that, we're saying that we want to have the circuit generate data that will be available for further processing. The bus we're talking about here is more than just the output of the keyboard circuit because we want to be able to latch the data. A latched, tristated data bus isn't something you get for free or by accident—it has to be planned.

3. All entered data will be displayed on LEDs.

Circuits that use keyboards usually have their own way of showing just what data has been entered. Computers, for example, use the keyboard output to decide what will show up on a video screen. Since there's really no way of knowing what we'll be controlling with this keyboard, there's no guarantee that any circuit using our keyboard will also have a handy way of displaying data.

There's another advantage to adding an LED display to the keyboard we're building—it makes the keyboard easier to debug. Instead of having to use probes to see if the keyboard is putting out data correctly, we can watch the display. If you finally use this circuit someplace where a display isn't needed, you can always leave it out. When you adapt one circuit for use in another, remember that

It's easier to leave stuff off than to add stuff on.

This rule is particularly true for the keyboard design because one of our goals is to design something that's as flexible as possible.

4. Each keypress will have an audible signal.

Some people call stuff like this nothing more than a piece of extra circuit glitz, but it's really a lot more than that. Having a signal generated to indicate that a key has been pressed is psychologically desirable for any kind of keyboard entry system. It's a simple way to let someone know that the keypress was recognized and that the data was entered successfully.

Having a beep sound whenever a key is pressed is also a nice thing to have while we're designing the circuit because it's a handy way of knowing that a portion of the circuit is working. How much it can tell you depends on what part of the circuit is generating the beep. The farther it's located from the keyboard switches, the more significant the sound is going to be.

It's much too early in the design (after all, we're just getting started) to plan the details of the beep generator or decide where in the final circuit we're going to put it. Once we start working on the block diagram, we can make some preliminary decisions. At this stage of the game, it's only important for us to know that we want to include it in the final design.

5. The keyboard will have two-key rollover.

One of the keyboard characteristics that people like to talk about is how lively the action is. This is something that depends on several things—the kind of switches used for the keypad, how quickly the data is entered, and so on. Individual tastes vary, but everyone who uses a keyboard makes a judgment about how responsive it is.

One element that measures the keyboard action is called *rollover*. This means that if more than one key is pressed at the same time, the circuit will process each keypress in turn. One of the best examples of rollover is a computer keyboard. Most computers have a keyboard buffer that stores a number of keystrokes. While this type of rollover is built with a combination of hardware and software, the principle is the same and so is the result.

Since we're only building a keyboard with 10 keys, there's no real reason to specify a huge number for rollover. Two-key rollover is a good goal for a circuit that will probably be used for the entry of numbers, control signals, and similar things. Computers need much larger buffers because they're frequently used for word processing and other applications where the rate of data entry is very high.

Once we finish the circuit, there's no reason why you can't enlarge the design to include more keys and a higher rollover. As a matter of fact, doing something like that is a great design exercise. For the moment, however, let's stick with the game plan. When you're just starting out, the best rule is

Keep it simple.

You'll understand the basic operation of the circuit better. Going out of your way to create potential hassles is not a smart thing to do.

6. The circuit will reset at power-up.

Circuit reset at power-up is a must for any data entry circuit. No matter what final use you make of this circuit (or any other based on it), it's a lot more than a convenience to know that the circuit will always be in a particular state when you first turn it on. It's also a good idea to have this state happen automatically because there's no way that we can be sure to remember to reset the circuit manually every time we turn it on. Even people with great minds like ours have been known to forget stuff (insignificant stuff, of course) every once in a while. Besides,

Brain space is valuable real estate.

Filling it with unnecessary stuff means there's less room left for things like junk food, old movies, great books, and other important items.

7. The circuit will have a manual reset.

Being able to clear the circuit from the keyboard itself is a standard requirement for any keyboard design. Everybody screws up occasionally when pressing buttons (except, I hope, the President of the United States), and needs some way to correct mistakes.

This is a fairly simple design decision, but how we'll implement it depends on the details of the circuit. Remember, we've already specified that we want the design to have several parts: a keypad circuit, a data bus, a display, and other things. To say that we're going to have some way to clear the circuit isn't all that definite until we know which parts of the circuit have to be cleared.

It's evident that we'll want to be able to reset the display to all zeros, but whether that will also clear the bus depends on the kind of plans we have for using this circuit when it's finished. Because this project is just an exercise in design, it really doesn't matter which way we go.

The best decision to make at the moment is to have the ability to reset everything so that the circuit is as flexible as possible and easier to adapt later on. When we finish this design, we'll know exactly how it works and be able to modify it to behave in any way we want for any purpose that comes to mind.

When you're designing general-purpose circuits,

Flexibility is the name of the game.

The point of doing all this work (aside from learning how to do it) is to have a circuit on the shelf that saves design time when you're working on other projects later.

8. All the parts used will be cheap and easily available in onesies.

Everything you do on the bench should be done with this consideration in mind, especially if you're working by yourself and (of even greater importance) you're the only one bankrolling the whole thing. I've always found it interesting that the projects described in books and magazines go into excruciating detail about everything from design philosophy to construction, but they never talk about getting oddball parts or cutting costs.

Think about it. When you're sitting down to design a circuit, it's no big deal to go through the databooks and find a chip that contains 90% of what you want and does 90% of your work for you. To a large extent, this is what professional design is all about. If you're in that situation, things are really great because you can get almost any part you want from the manufacturer for just the price of a phone call. They call this stuff *engineering samples.*

It sounds great but the reason the manufacturer goes out of his way to offer samples is that they might lead to a really massive order—for honest-to-God money—a bit down the road. There's nothing stopping you, sitting alone at your bench, from trying the same thing; but it takes a lot more than a basic desire for the part to shake a few of them loose from the manufacturer. Chances are you'll be directed to distributors. Call them and you'll learn that while you can certainly buy one or more of anything they have, they're "sorry, but there's a minimum order of XXX dollars."

Even if you're able to get a couple of freebies, don't forget that you're going to be using them to design a prototype circuit, not plug them into a socket on a completed design. As we all know, funny things have a habit of happening (all too frequently) when you're working out something on the bench.

One of the most important rules to keep in mind when you're in the middle of circuit development is

Don't use what you can't replace.

That's true regardless of whether there's a problem because of cost, availability, or anything else.

There's nothing worse than having a part go up in smoke because of a stupid mistake and then being forced to wait weeks for a replacement to arrive. Be smart.

Once you've worked out all the details and have a final design in front of you, you can start thinking about how to shrink it down with the use of special-purpose silicon. Until you're sure that you've reached that point, however, chip count and circuit complexity are totally irrelevant.

First make it work, then make it neat.

A poorly designed circuit that works is a lot more impressive and valuable than a slick one that doesn't.

Block diagram—fleshing it out

Now that we know what we're trying to do, the next step in any design is figuring out how we're going to do it. The list of eight design criteria we've decided on is the basic guide for the circuit we'll be developing, but it's a long way from something we can plug into the wall.

If you look at the anatomy of the design process—and I'm talking about the whole thing now—from a vague stirring in the back of your brain to a working reality on every retail shelf in the universe with the exception of Neptune (since there's nobody there anymore), each step in the process is designed to lead to and help you with the next one.

Once you're looking at a list of the things you want the circuit to do, you've got to analyze it and see what kind of electronics will be needed. This part of the process usually happens quickly because a lot of the needed thought has already kicked around in the back of your brain while you were working out the criteria.

Creating a block diagram is probably the most crucial part of the entire design. All parts in the design process are important, but this one is where you get an idea of the amount of work ahead of you. You've got to work out the individual sections of the complete circuit and figure out how they have to interact to get the job done.

Identifying the parts

The first step is to identify each part of the circuit. Once that's done you can see how the parts have to be connected. Also, you can get a handle on any potential problems to watch out for or special considerations you might have overlooked up to now. Breaking the job into pieces like this means you'll be able to tackle each one as a separate design project. In addition, you'll be able to cut the original design job down into bite-sized chunks and make the best use of your time.

Our project, a keyboard data entry system, isn't really up there with stuff like a variable interossiter or a Krell mind expander (and 12 gold stars if you know what those are), but every project benefits from being approached systematically and

Logical thinking produces logical circuits.

Regardless of size or complexity. I mean, if you don't have a clear idea of what goes into a design at the beginning, how can you possibly expect to be able to put it together at the end?

Since the circuit we have in mind only has one purpose—data entry from a keyboard—we don't have a huge number of separate sections to design. That's not to say that we don't need a block diagram; ours just won't be very complicated.

To start off, one of the most basic sections we'll have to design is the keyboard. We specified that we want to be able to enter up to 10 separate digits; thus, we'll need a circuit that can accept 10 switches and indicate which of them has been pressed. Specific design considerations come later. What we're into now is working out exactly what we have to design in the first place.

The first box in our block diagram, therefore, is going to be the keypad and its associated circuitry. The keypad is the first section because it's right at the beginning of the circuit. It's sometimes hard to figure out where to start in drawing a block diagram. After you've done a few of them, you'll find you have your own preferences. I like to begin with one end and work from there, but other designers

I know like to start at the power supply, the output, or any other section they can immediately identify. (They're all wrong, of course.)

Until you get a bit of experience behind you, the best way to lay out the problem is to start at the inputs and work your way toward the outputs. This method may take a bit longer to finish, but it's easier and much more logical. Also, you're less likely to overlook something.

Now that we've identified the keypad, it follows that the signal coming out of it has to go somewhere—it has to serve as the input for something else. In our case, all the keyboard is doing is telling us which of several keys has been pressed. We've got to take that information and use it to produce data in some usable form. Although we really haven't talked a lot about the kind of data we want our circuit to generate (and you're free to make up your own mind), probably the most useful data we can generate is plain old binary—or hex if you like to think of it like that.

A small aside:

Question: Why do the fabled fish-faced people of Neptune use computers based on the number 10?

Answer: Because they have 16 fingers.

So maybe you had to be there.

Whatever you think of that—hey, I think it's funny—it's clear that we have to add a binary encoder to the circuit. Because we're only dealing with 10 digits, we can further specify that it's going to be a binary-coded decimal (BCD) encoder. If, at a later time, you want to expand the circuit to handle all 16 hex digits, you'll only have to make slight changes to the keypad and the encoder. We're keeping things flexible, remember?

Sometimes the criteria you specify for a design also specify one of the elements of the block diagram. We indicated that we wanted the circuit to beep every time we pressed a key on the pad; thus, we need a circuit to generate the beep. The third item in the block diagram will be a beep generator. (All things should be so simple.)

We also wanted all the entered data to show up on a data bus so, as you would guess, we have to design one; and that becomes yet one more item in the diagram.

We also specified that we wanted the data to show up on a display, and that means we have to add this as well. It's not necessary for us to indicate what kind of display (LED, LCD, etc.) because this is a detail we'll work out when we attack that part of the design. All we have to know at this stage is that we're going to have to work out the design for one before we can finish the project.

There are still two more sections that we have to add to the block diagram, and I've saved them for last because they're not immediately obvious. I said that you should populate your block diagram from the inputs to the outputs and that's still true—but sometimes that makes you forget about things.

Because we're using one keypad to generate up to a 10-digit number, we need some way to have the circuit sequence our keyboard output from the first digit to the second digit, and so on, through to the tenth digit. This part of the final circuit has to keep track of the number of entries and make sure they all stay in the correct order. For want of a better term, let's call this the digit selector and add it to the list.

The last item could be considered as a built-in part of a different section of the block diagram, but no harm can come by listing it as a separate section. Although we aren't dealing with the details of any section in the block diagram (and we won't until we start designing hardware), we can't forget about them entirely either.

It's probably true that however we design the keyboard, the keys themselves will be made with mechanical switches. Now whenever you use mechanical contacts, you have to make sure that one press on the pad results in one, and only one, signal to the rest of the keyboard circuit. All mechanical switches have to be debounced if you want to be able to use them reliably. Most LSI keyboard encoder chips do this internally; but since we're going to be designing ours from the ground up, we have to make it a definite point to include circuitry to clean up the keyboard signals. The last box in the block diagram will be the switch debouncer.

And so, believe it or not, we've reached the end of things we have to include in the block diagram.

There are other things we could have added to the list, such as the power supply, but the more things you add, the more confusing the job can seem to be. Whenever you get to the point of drawing up a block diagram, you have to strike a balance between specifying the essential sections and overpopulation. In the case of a power supply, all we'll be needing is a simple 5-V supply (or even something as basic as a battery), so there's no need to blow things out of proportion by listing it separately in either the design criteria or the block diagram.

If the project you were designing required a more complex supply, it would certainly be listed on its own. Ours is a really clear-cut example but, I'm sad to tell you, electronics is just like life because

Nothing is ever clear cut.

You're the only one who can decide if a particular element of your project has to be individually specified and separately designed. One good rule of thumb is that if it's not a major part of the overall design and you know exactly how you're going to lay it out, you can leave it out. But if you have any doubt whatsoever, leave it in.

We now know the elements that have to go into our keyboard project. If you've been keeping track, you know that there are seven items in the block diagram:

- Keypad
- BCD encoder
- Switch debouncer
- Beep generator
- Digit selector
- Data bus
- Display

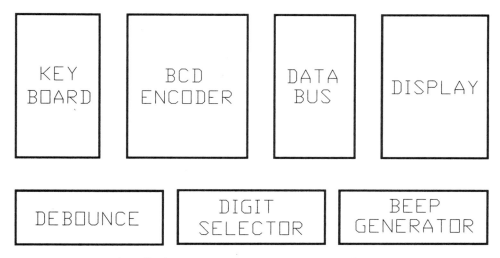

3-3 Basic elements of the keyboard block diagram.

While I've got them all drawn schematically in Fig. 3-3, it's far from the end of the story. Now that we know what components we have to design, we have to figure out how to connect them and make them work as a keyboard data entry system. After all, that was the original plan . . . remember?

Connecting the sections

There's a limit to the amount of detail that's needed or even desirable when you're doing a block diagram. While it's important to list all the sections that make up the final circuit, too much detail can clutter things to the point where the diagram is so confusing it becomes totally useless.

So how much detail is enough? Where do you draw the line, so to speak?

I know you expect me to give you some sort of hard-and-fast rule to follow that will tell you when you're done with a block diagram. If you know one, I'd like to hear about it. I've seen diagrams that ran the gamut from just a few lines and boxes to things that weren't much less detailed than the final schematic. Since the whole idea behind a block diagram is to serve as a guide for the development of the circuit, the amount of detail depends completely—let me repeat, completely—on the kind of guide you need. And you're the only person in the universe who knows what you need.

If you're relatively new to electronic design or the project you set for yourself is extremely complex—in other words, there's a good reason for a detailed block diagram—then that's what you should have. More experienced developers or people working on a circuit that's based on previously designed sections will need less information.

The bottom line is to develop only what you need.

Extra work is wasted time.

If you have any doubts, it's better to err on the safe side and put in the extra details.

Now that all the individual sections have been identified, you have to give the same care and concern to showing the signals that will be going from one part of the final circuit to another. This isn't too hard to do, and just about the only thing to watch out for is adding too much to the diagram. All you want to do here is show signal paths and signal flow, not each wire that you're going to need when you sit down to build the circuit.

The best way to understand the kinds of things you should indicate in the diagram is to look at an example and see what's been left out of it. Figure 3-4 is a good block diagram. As you can see, all the sections we defined earlier are included, and the main signal paths have been drawn in. I've used arrows because it's useful to know not only what signals are running around the circuit, but also what directions they're traveling in. The lines drawn in our block diagram are all single-directional arrows; but if any of them had been bidirectional, the lines would have had arrow heads on each end.

The most interesting thing about the diagram is the information that hasn't been included. You'll notice, for example, that power and ground aren't shown. They've been left out because there's nothing special about the power requirements for the circuit, and drawing them in wouldn't make the diagram any more useful. As a matter of fact, it would only clutter things up and make it more difficult to see how the circuit is laid out. If the power supply had been a major design consideration, it probably would have been necessary to draw in the paths for the various voltages, feedback lines, and anything else that would deserve special attention.

Before we start talking about the block diagram in Fig. 3-4, there's one more thing you should notice. Leaving both the power and ground lines off the block diagram is a pretty obvious thing to do, but there are also other things we left out that may not be so immediately obvious.

If you look back over the list of criteria we specified for the design of this circuit, you'll notice that we made a definite point to include both automatic and man-

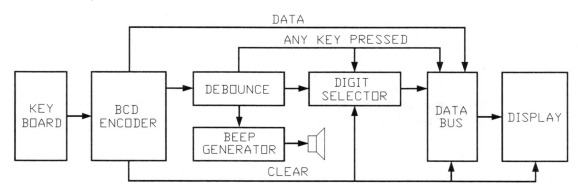

3-4   Complete block diagram for the keyboard project.

ual reset. The block diagram shows a reset line going from the keypad to the data bus. This is the signal to manually clear the data bus and display so I drew it as an indication that it only has to control those two sections.

What isn't shown in the diagram is the automatic reset that takes place when the circuit is first turned on. While this specification is important enough to list as a separate design criterion, there's no reason to include it in the block diagram because it's not a separate circuit by itself. Power-on reset is a consideration for the keypad circuit, the BCD encoder, and probably the data bus as well. When we get to the design of each of these sections, we'll incorporate it in the circuit.

What's important to realize here is that when you're planning the project and working out the overview, you have to be able to recognize the difference between an individual section of the design (something that shows up in the block diagram) and things that are part of each section. Remember that the reason for doing a block diagram in the first place is to generate a drawing that will provide an overview of the final project, not to indicate every component that goes into building it. When you start adding resistors, capacitors, and other board-level components, you're talking about a schematic, not a block diagram.

Before we get started talking about specific hardware and begin the actual design of the project, there's one more important point to make about the planning stages.

Up until now, we've only been using four tools: paper, pencil, eraser, and the stuff between your ears. Mistakes can be corrected with the eraser, and changes can be done with the pencil. While these are easy things to do, the majority of the work is being done by your brain. As you flesh out the design criteria and block diagram, you're setting the amount of work you'll have to do later on when you get into the nitty-gritty of actual circuit design.

The better the job you do in the planning stage, the easier you'll have it when you start translating the block diagram into the working circuit. There's always a temptation to cut the paperwork short and get into the hardware, but that's not the way things usually turn out. The better the paperwork you bring to the bench, the faster and more hassle-free it is to deal with circuit design.

Regardless of whether you're working on a project of your own or doing a design job for someone else, the answer to anyone (including yourself) who wants to hurry things up is always

Do you want it Thursday or do you want it good?

Spending less time generating paperwork is exactly the wrong way to save time. If you start doing hardware without both good notes and a carefully thought out block diagram, the design phase of the project is going to take more time, create more hassles, generate excess brain damage, and, when you ultimately get finished, the final product isn't going to be as good as it could have been. Guaranteed.

But that's not a problem here because we're being systematic almost to the point of tediousness. Now that we've reached the end of the planning stage and have a handle on the work ahead of us, the next step in the project is to get our hands dirty by starting the hardware phase of the job.

That's right, now that we know what the circuit has to do and what sections we need to do it, all we have left in front of us is figuring out how to do it.

A slick trick

Need a little box that can produce an audio output with a variable gain and frequency? I built this handy circuit a few years ago and have found it invaluable for testing microphones, telephone cable, and just about any place where you need an audible frequency.

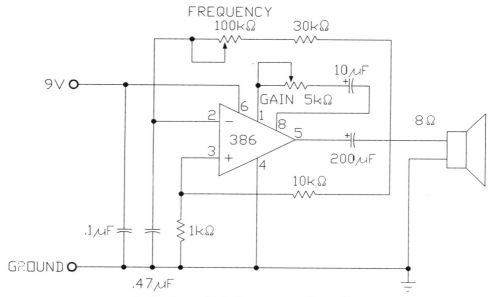

A simple, variable-frequency audio oscillator.

The circuit is super reliable, and I always take it along with me when I go out on a job because it's tiny enough to fit in the bottom of my bag and I never know when it's going to be needed. There's nothing critical to keep in mind when building it, and it works happily when powered by a 9-V battery.

4
Keyboard design #1
Breaking ground

Everybody has a method for dividing people into groups. There are lots of standards. It's true that different things are important to different people. My dentist told me that when he was in school one of his teachers told him that the job of the human heart was to supply blood to the gums.

When I was a kid, I always divided the food on my plate into good stuff and bad stuff. This division didn't have anything to do with the nutritive value of the food, of course; it was based solely on how the food tasted to me. I always ate the bad stuff first and saved the good stuff for the end. I still do the same thing.

When I first started to do electronic design, I used to work as a member of a large team. Because things were usually organized, I was responsible for only one small part of the entire job. The more I worked like that, the more I felt that I was learning how to be the world's foremost expert on something that would more than likely be obsolete in a week.

Having the responsibility for an entire project—from the paperwork to the prototype—is the single most exciting and rewarding thing you can do. But not everybody likes doing every part of a design. Human nature being what it is, there's a real tendency to shortchange the least enjoyable and drag out the most enjoyable. That's a really big mistake.

If you're the kind of person who likes doing empirical design right at the beginning of the project, you're letting yourself in for a lot of trouble. The same is true for paperwork freaks who find that they like drawing lines a lot more than working with real hardware at the bench. Each step of the design process is equally important, and cutting any step short can only cut down the success rate you'll have for all the projects you start.

Completing a project means taking care of both the good stuff and the bad stuff. You just can't cut corners. A lot of companies have spent a lot of money and made a lot of mistakes that cost a lot of money to finally arrive at the best procedure to follow for research and development. You're free to do whatever you want, but if you want to be as successful as possible, you should profit from other people's mistakes and remember that:

First you think and then you do.

Action without thought is . . . well, it's thoughtless.

Starting the design

The roadmap for the circuit we're going to build is the block diagram we drew up in the last chapter. Figure 4-1 shows the individual elements that have to be turned into reality and indicates how all the sections have to be connected.

Before we get into working on the hardware, take a good look at the block diagram. This is the last time we'll be concentrating most of our attention on the design of the overall project. One of the main reasons for doing all the work we did in the last chapter was to give us the ability to concentrate on designing one part of the project at a time. While our goal—designing a keyboard data entry system—is the completion of a relatively small project, we can all still benefit from breaking it down to smaller, bite-sized pieces.

Once you get into the habit of analyzing the job you set out for yourself, you'll find that even the most complex project you can think of is just as manageable as the simplest. Circuits whose schematics take up a whole book aren't meant to be understood in their entirety as if they were novels. You work on understanding one section at a time, not everything all at once.

The paperwork for the electronics in a modern fighter plane covers volumes; and nobody—repeat, nobody—has an off-the-top-of-the-head understanding of all of it. Nobody remembers every detail. A friend of mine was on the electronics design team for the F-15, and I doubt that you'll find anyone more familiar with the ins and outs of the circuitry in that plane.

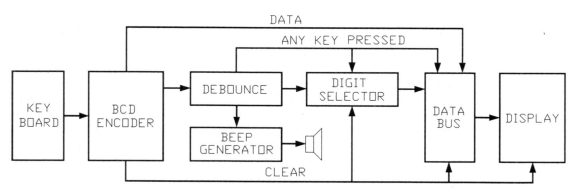

4-1   Block diagram of the keyboard.

Even though he's been through the electronics over and over again, he still has to refer to the charts and schematics when he works on it. What he has is an understanding of the overview of the design, not total mental recall of all the details. When he starts to troubleshoot the system, the first piece of paperwork he pulls out is the block diagram, not the schematics.

When he locates the section he wants to work on, he'll go to the schematics and get the paperwork for that section. While he's working on that section, he treats it as an individual circuit. This means checking input and output signals, internal voltages, clock pulses, and so on.

Even though you probably won't find the project we're working on in the cockpit of an F-15, we'll be doing the same sort of thing. We may be going into more detail, but the procedures we'll be following are the same ones used on every project regardless of the size, complexity, or intended use. Every single design— and that means everything from the electric knife in your kitchen to the space shuttle at Edwards Air Force Base—starts out life by being broken down into pieces small enough to be completely understood by a single human being.

The method is always the same: no matter what you come up with, no matter how big or how small it is,

First you take it apart, then you put it together.

That's the only way you'll ever get the whole job done.

Designing the keyboard

It seems that I've been spending a good deal of time so far telling you about important parts of the design process and warning you not to forget one thing or another. Well, when the time comes to get into the hardware part of things, there's one more important point that you should keep in mind. It's even important enough to write down and hang on the wall:

Don't even think about hardware without having a databook.

There are no words strong enough to use when it comes to stressing the importance of databooks or sheets. ICs may look like simple little things, but there's a lot of stuff going on down there on the substrate; and unless you know about it, you're leaving yourself wide open for potential disasters (good things to avoid).

The more exotic the parts you're going to use, the more important it is to have a databook for the parts. One of the design criteria we drew up earlier for our project was that we were going to use parts that are cheap and easily available so you'll find it relatively easy to get data. No matter how familiar you are with the parts, however, you should make it a point to have the information around anyway.

Keyboard circuits come in all varieties, but most of them basically decode the output of a running counter. The counter continuously and sequentially puts out all

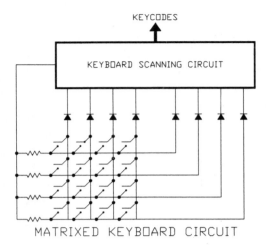
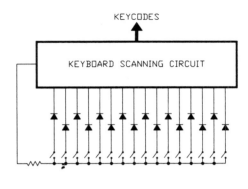

4-2   Matrixed and common-leg keyboard arrangements.

the codes you want the keyboard to generate. By pressing one of the keyboard switches, you stop the counter and the code associated with that keypress remains on the bus.

There are two basic approaches to designing a keyboard circuit that scans through a sequence of output codes. The difference between the two is in how the key switches are connected to the circuit. Both methods are shown in Fig. 4-2.

In a matrixed keyboard, the switches are arranged in an XY matrix so that pressing one key causes two of the circuit lines to be connected. This configuration is fairly common in computer keyboards—once upon a time, it was just about the only one you could find. When a key was pressed, the silicon in the computer would decode the state of the X and Y lines (a series of ones and zeros) and translate that into something that the computer could understand. Some computers wanted the keypresses translated into ASCII, while others (like the Apple) had their own special codes.

The advantage of matrixed keyboards is that you can handle a lot of keys with relatively few lines. The disadvantage is that, unless you use special silicon, the circuitry can get to be a bit complicated and the PC board patterns can be a nightmare. If, however, you need a keyboard with lots of keys on it, this is the way to go.

Common-leg keyboards have the switches wired so that all the switches have one leg in common (hence the name). As you can see from the illustration, the biggest advantage of this type of arrangement is that it's much simpler. The disadvantage is that if you get much beyond 15 switches or so, the design starts to get awkward and the amount of circuitry needed to support it gets rapidly out of hand. This type of keyboard is the kind you find in calculators, some telephones, and other applications where there's no need for lots of keys.

Since we've already decided that we only want 10 different keys on the keypad, it makes good sense to go with the common-leg approach and that's exactly what we're going to do.

There are as many ways to get a job done as there are jobs to do, and nowhere is that statement more true than in electronics. Designing a common-leg keyboard is a simple exercise in logic, and I'd be surprised if you didn't have a few ideas in mind right now. Since the only real constraint we've put on ourselves is to keep the parts available and cheap, we'll keep this near the top of our priorities.

If you think about the problem for a moment, you'll realize that three separate things are needed for this design (in no particular order):

- Keyboard clock
- Keyswitch decoder
- Keycode encoder

Since we're just starting out on this design and there's no way for me to know how much experience you have had, we'll go through everything step by step. But don't worry. By the time we get to the end of this book, you'll all be well on the way to becoming design experts. Guaranteed. Trust me.

Keyboard clock

There aren't many special requirements to keep in mind when you have to design a clock that's going to be used in a keyboard. Just about the only consideration is that the frequency should be kept below the five-digit range. This is sort of a Goldilocks consideration: too low a frequency causes a sluggish response at the keypad, and too high a frequency can cause output errors. Some numbers are too high, some are too low, and some are just right.

The best all-around choice for a clock generator is a chip that you more than likely have used lots of times before—the 555. Even though you can generate frequencies with something as simple as a single gate or transistor, the 555 is such a convenient and easy chip to use that there's no real reason to look anywhere else. Let's not forget that one of the most important rules to keep uppermost in mind when you're working on the bench is:

Avoid hassles.

You'll undoubtedly run across enough unexpected problems without going out of your way to manufacture any.

Even though you should have databooks, I'll make it a practice to give you enough information about each chip we use to get you through the design we're working on. While we're on the subject, I'll also list the equipment you'll need to find out if the circuit is working properly. In the beginning you can get by with a multimeter and a logic probe; but as things get more complex, you'll need more than that.

The ultimate piece of equipment is an oscilloscope. If you already have one, you can use it in place of just about everything else. At this stage of the game an oscilloscope is overkill, but it certainly makes things a lot easier. I'll let you know when a test can only be done with a scope, and I'll always have pictures of any critical scope stuff.

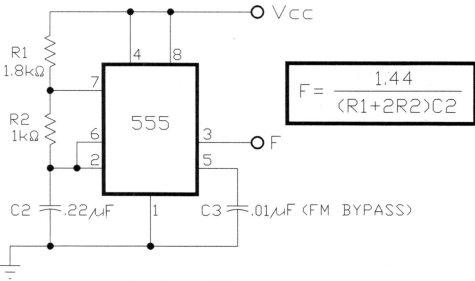

4-3 Using the 555 as a clock generator.

The basic information about the 555 is shown in Fig. 4-3. Keep in mind that the information there only relates to how to set up the chip for use as a frequency generator (or astable multivibrator if you prefer being obscure). With the components shown, the output of the circuit is about 2 kHz—a good Goldilocks choice for our needs.

The only special thing to notice about the circuit, and one you may be wondering about if you've never seen it before, is why capacitor C3 has been hung on pin 5. The 555 has gotten a reputation of having an output frequency that's wonderfully immune to problems and noise that come from less than perfect power supplies, but there are still times when dirty power can cause screwups at the output.

If you're interested in finding out this sort of stuff, I'll repeat my earlier suggestion that you get yourself a databook. As far as this circuit goes, take my word for it that C3 can eliminate lots of hassles. The 555 will probably behave like a real trouper without it, but capacitors are cheap; and bypassing the FM input like this is a good habit to get into. It doesn't hurt and, who knows, it might help.

Keyswitch decoder

Building a keyswitch decoder isn't all that difficult. One reason is that we've almost defined the part to use in the list of design criteria we drew up earlier. Even though we weren't talking about hardware details then, a glance at some of the specifications should give you a clue to what I have in mind.

Since we're designing a BCD encoding keyboard, we'll be having 10 switches on the keypad. We've already decided that because of the relatively small number of switches, it's a good idea to use a common-leg keyboard design rather than a

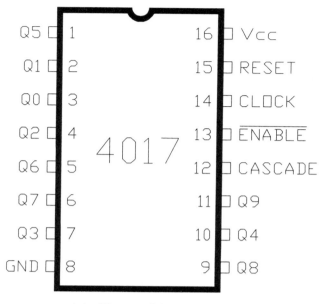

4-4 Pinouts of the 4017.

matrixed one. This is starting to sound as hokey as "Wheel of Fortune," so take a look at Fig. 4-4 and see what I'm talking about.

The 4017 is a 10-step Johnson counter that's absolutely ideal for our purposes. This chip was a big hit when it first appeared, and in no time at all it became a certifiable silicon superstar—the Barbra Streisand of circuit designers. You should have a databook in front of you but, just in case, let's go through the pins one at a time.

Outputs   The outputs are on pins 1 through 7 and pins 9 through 11. These pins are normally low; and each goes high, in turn, on incoming low-to-high transitions of the input clock. As with a lot of chips that have sequential outputs, the pin order has nothing whatever to do with the output order. I've been told by a friend in the chip design business that establishing a logical order between pin numbers and outputs isn't a priority item. Maybe if we all signed a petition . . . probably not.

Cascade output   The cascade output is on pin 12 and goes through one complete high-to-low cycle each time the outputs cycle from 1 through 10. This pin is high for counts 0 to 4 and low for counts 5 through 9. Although this output is usually used to either get a frequency that's one-tenth the input frequency or as the clock for another chip, it's important to realize that the 4017 has to go through a full count of 10 for the output to work. It's tied to the 10 outputs, not the input clock. This means that if your circuit resets the 4017 to zero whenever it reaches a count of less than 5, the cascade output will always remain high.

Enable input   The enable input, pin 13, controls the output sequencing. Putting a low on this pin will let the chip count normally; making it high will freeze the state of the output pins, and the chip will ignore the input clock. Remember that there's a difference between disabling the count and disabling the chip. One thing to keep uppermost in mind when you're using a 4017 is that there's no way to make all the outputs low—you can't have all the outputs turned off at the same time.

Clock input   The clock input is on pin 14. This is a standard input with standard requirements. The more glitch-free and squarer the clock, the happier the chip will be. Because this is a CMOS chip, you can be pretty sloppy with clock pulses and still have the chip work well. The count will advance from output to output with each negative-to-positive transition of the input clock.

Reset input   The reset input is on pin 15; and making it positive will cause the 4017 to put a high on pin 3, the first output. (Remember what I said about pin and output numbering?) As long as the reset input is held high, the chip will ignore incoming clock pulses and keep pin 3 high. Once reset is made low, the chip will start counting again.

Other pins   The other pins are 16 and 8, the power and ground inputs, respectively. These are both standard stuff; and since we're dealing with a CMOS part, the power standards are hassle-free. We'll be operating the circuit off a 5-V supply, but you can use anything up to 12 V or so without any problems. When you're in the breadboarding phase, it's a good idea to use a voltage below the maximum allowed. We all know that

Things do happen.

And it would be a real shame to be two weeks into the project and have everything go up in smoke because the power supply happened to sneeze a bit. (Maybe *shame* isn't the right word.)

Keycode encoder

The keycode encoder is the part of the circuit that takes the keypress data you enter and converts it into some kind of code. In our case, we're going to translate things into BCD, but that's not the only way to go. There are as many encoding standards as there are needs for them, and a lot of equipment using keyboards outputs codes unique to their own applications.

Over the last few years, I've gotten a lot of mail from people who have used some far-out methods to generate output codes. Some of these are really clever, and some say more about the mind of the designer than the efficiency of the design. But no matter how you go about it, the best way to handle the problem is to find a part that will do most of the work for you. Because we've decided to use cheap, available parts, the next step is to flip through the databooks. The data we want to produce is standard BCD; thus, it should come as no surprise that there are standard parts for doing the job. One of these, the 4518 BCD counter, is not only exactly what we're looking for; it has some extra stuff that might come in handy later on as we get further into the design.

While you're flipping pages in the book, you'll probably come across chips that can take care of the whole keyboard design. They might be a bit more expensive and harder to get in onesies, but it's reasonable to look at them because they'd take care of everything we talked about doing. If the only point of going through this design exercise was to come up with a working circuit, a keyboard chip would be the way to go about it.

But we're doing more than that. Remember, we're not only learning how to design one particular circuit; we're learning how to design any circuit. Even though we're paying attention to the specifics of the electronics, what's much more important is the method we're following. Whenever you sit down at the bench—no matter what project you're working on—you have to clear your mind, concentrate on the problem, and

Think straight about stuff.

Or you'll be wasting lots of time, and the end product—no matter what you call it—will probably do nothing more than drain batteries.

The 4518 is a dual BCD counter—two independent packages in the same chip. A databook will tell you everything you could possibly want to know about the chip, but let's go over it in outline form (refer to Fig. 4-5).

Although the pinouts make it look very different from the 4017 we talked about earlier, the chips are very similar. Both of them are synchronous decade counters, and the main difference is only in the way they count. While the 4017

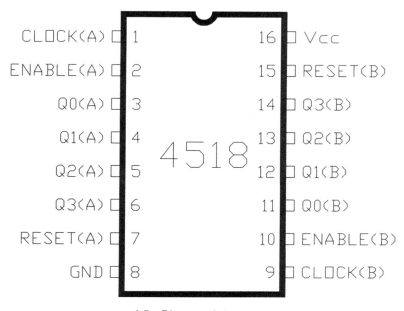

4-5 Pinouts of the 4518.

needs 10 separate outputs, one for each of the counts, the 4518 only needs four because it counts in binary rather than decimal.

The 4518 is a *synchronous counter*; that is, all the outputs change in response to the pulses on the chip's clock input. In other types of counters, only the first output is triggered by the clock input. In this design the first output clocks the second output, the second clocks the third, and so on. These counters are often referred to as *ripple counters* because incoming clock pulses cause the count to ripple through the chip.

Synchronous counters have outputs that always change state in step with the input clock. The main advantage of doing things this way is that there's always a valid count on the outputs because they all change state at the same time.

The only synchronous output in a ripple counter is the first one because that's the only one that's being driven by the input clock. All the other outputs in the chain use the preceding output as a clock. Ripple counters are simpler to build, and you can have lots of outputs on one chip. All of the CMOS long-chain counter chips—the 4020, 4040, 4060, and so on—are ripple counters. This may seem a terrific way to build these things because you can pack a lot of arithmetic on one substrate. Before you undergo a conversion to being a true believer, however, remember that

There's no such thing as a free lunch.

Everything has a price. It may not be immediately obvious—it may even be hidden—but you can be sure it's there.

The downside of a ripple counter is that there's no way to be sure that there's always a valid count at the outputs. This can be a real problem. Whenever you have your circuit glance at the counter to read the state of the outputs, there's a good chance the chip will be lying to you.

The concept of silicon dishonesty may be an electronic eye-opener, but that's the way it is. You can guard against this but—hey, if you can't trust electrons, what's the point of anything?

The easiest way to understand the concept is to go through an example. As a ripple counter goes from a count of, say, 7 to a count of 8, funny things happen at the outputs. Remember that each output uses the previous one as a clock.

$$\begin{array}{ll}
\text{The truth} & -0\ 1\ 1\ 1-\text{``7''} \\
\text{Lie No. 1} & -0\ 1\ 1\ 0-\text{``6''} \\
\text{Lie No. 2} & -0\ 1\ 0\ 0-\text{``4''} \\
\text{The truth} & -1\ 0\ 0\ 0-\text{``8''}
\end{array}$$

Now it's true that the invalid counts are only there momentarily; but considering how things usually work in life, those would probably be the times you wanted to read the outputs. The circuit hassles caused by something like this would be murder to track down.

Historians have determined (from examining recently returned pictures by *Voyager*) that transitory invalid counts on the outputs of a ripple counter are what led to the extinction of the fabulous fish-faced people of Neptune.

It's a little-known scientific fact that the prehistoric Earth was used by the Neptunians as a vacation resort. Newly excavated papers show that the last great Neptunian scientist, Klant Zorch, warned that basing the construction of Atlantis (intended to be the Disneyland of Neptune) on the data from ripple counters would result in absolute disaster. You can't argue with history.

Using a ripple counter for encoding circuits is a good way to leave yourself open to considerable brain damage. The counter's main use is really keeping track of how many events occur over a period of time.

Because we want to use a counter as an encoder, we have to be sure that every time we press a key on the keypad there will be a valid number at the output of the encoder. That means we have to use a synchronous counter. The 4518, our chip of choice, is made of two identical counters; thus we'll only go through the pinouts of one of them.

Outputs   The outputs of the first section are pins 3 through 6. These are binary outputs and are weighted in a binary 1, 2, 4, 8 sequence. The 4518 is a BCD counter so the count will go from a binary 0 (0000) to a binary 9 (1001) and then recycle to 0. Because this chip is a synchronous counter, we don't have to worry about invalid data showing up at the outputs. You can snatch the data off the outputs whenever the mood strikes you and be secure in the knowledge that the data you find there is the data that's supposed to be there.

Reset input   The reset input, pin 7, is held low during counting operations. A high on this input will cause the counter to reset and put a zero on the outputs. As long as reset is held high, the counter will remain at zero and ignore incoming clock pulses.

Clock input   The clock input is on pin 1. Even though CMOS chips are pretty relaxed about the quality of clock pulses, the cleaner the waveform, the happier the chip will be. This means the input clock should be as square as possible and the rise time of the pulses should be no longer than about 10 μs. The 555-based clock we're using produces clock pulses ideal for the 4518. The count will advance one step with each positive-going clock pulse.

Enable input   The enable input on pin 2 is held high when you want the chip to count the pulses coming into the clock input. If you ground this pin, the 4518 will ignore any incoming clock pulses and the output pins will hold the count that was there at the moment the enable pin was grounded. Note that this pin disables the clock but leaves the output state unchanged. This is very different from the reset pin because putting a high on the reset pin will both disable the count and reset the outputs to zero.

Other pins   The other pins are 8 and 16, which are the ground and power pins, respectively, for the chip.

Before we leave the setup in the 4518, it's important to clear up a possible source of confusion over the clock and enable inputs. It's a bit misleading to call one a clock and the other an enable. Both pins are really clock inputs; and if you look at Fig. 4-6, you'll see what I mean. This is a peek at the innards of the 4518, and you can see that the only real difference between the clock and enable inputs is that the clock input goes through an inverter.

Both inputs are connected to the input legs of a NAND gate, and it's the output of this gate that actually controls if and when the counter will advance. If you're like me, you'll find this stuff fascinating; but if you're only interested in the more practical side of things, you can work out the truth table (an honorary membership in the elite Neptunian Guard for doing it yourself) or sneak a look at the truth table in the illustration.

The truth is that both inputs can be thought of as a clock and an enable; which is which depends entirely on how you want to use the chip and how you wire it in the circuit. Hold pin 2 high and the counter will advance with each positive-going pulse on pin 1. Hold pin 1 low and the counter will advance with each negative-going pulse on pin 2.

Some conditions, shown in the truth table, won't allow the counter to advance even if the reset pin is held low. We'll be taking care of this in our keyboard circuit; but when you use this chip in your own designs, you have to be sure to arrange things so that your circuit doesn't inadvertently miss counting one of the clock pulses. You don't want to have to track down a problem like that.

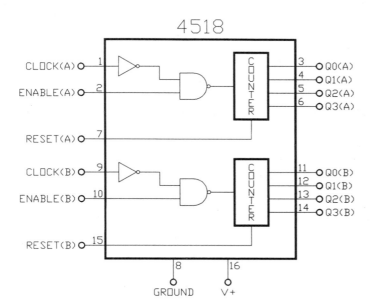

THE 4518 TRUTH TABLE			
CLOCK	ENABLE	RESET	OUTPUTS
⌐	HIGH	LOW	ADVANCE
LOW	⌐	LOW	ADVANCE
⌐	X	LOW	NO CHANGE
X	⌐	LOW	NO CHANGE
⌐	LOW	LOW	NO CHANGE
HIGH	⌐	LOW	NO CHANGE
X	X	HIGH	LOW

4-6 Block diagram of the 4518.

Wiring things up

Believe it or not, the time has come to actually put ICs and stuff on the breadboard, apply power, and hope for the best. We all know that the chance of having everything work properly the first time is about the same as finding a McDonald's on Neptune (delivery costs made the prices too high); thus, you should have some test equipment on hand to trace any problems that arise.

For this stage of the circuit, you can get by with just a multimeter and a logic probe. As the circuit gets more complex, you'll be needing more complex equipment—but we'll worry about that later.

The circuit we're putting together is shown in Fig. 4-7, and the layout (if you're using the same solderless breadboards I am) is in Fig. 4-8. We've already talked about a lot of the stuff shown in the schematic (particularly the 555), so we'll concentrate on how we're using the 4017 (keyswitch decoder) and the 4518 (keycode encoder) in the circuit.

The clock pulses from the 555 (IC2) are fed to the 4017 as well as both halves of the 4518. We're only using the first section of the 4518; but since this is a CMOS chip, we can't let any of the inputs float or the operation of the whole chip will be affected. You can connect them (the inputs of the second section) to power or ground; but as a matter of personal preference, I like piggybacking unused inputs to real board signals. If you're looking for the reasons why I do this, one good one is that it gives me a chance to see whether the other half of the chip actually works.

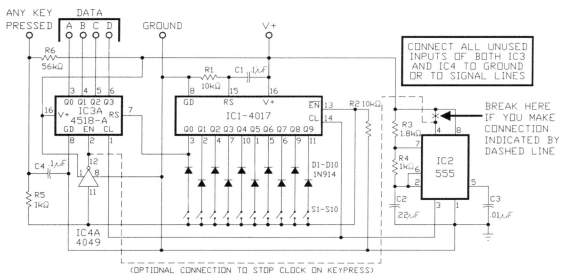

4-7 Schematic of the initial keyboard circuit.

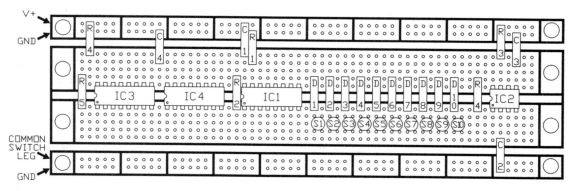

4-8   Placement diagram of the keyboard circuit.

My other reasons are due to things buried in my childhood. But on to things of consequence.

The 4017 (IC1) has its outputs connected to the 10 switches on the keyboard. The CLEAR switch specified in the list of design criteria will come later. We don't have to worry about that until we get into the design of the rest of the circuit. The common leg of the switches is connected to the chip enable input at pin 13. This line is held low (enabling the 4017) through R2 (whose other end is connected to ground). As long as no keyswitch is closed, the 4017 remains enabled and its outputs go high, in turn, at a rate determined by the frequency of the clock from the 555.

When a keyswitch is closed, nothing happens because the outputs of the 4017 are normally low and only go high when they're selected by the chip's internal sequencer. As soon as the output is selected, the high signal is routed through the switch to the enable input and the 4017 stops counting. If a second key is pressed at the same time, the count will stop there as soon as the first key is released. This is what's referred to as *two-key rollover*—one of the things we specified in the design criteria.

While we're looking at this part of the circuit, you should be able to figure out why there are diodes sitting between each of the outputs and the keyswitches. You probably know already but, just in case, the diodes are there to isolate each switch. When two switches are thrown at the same time, a high and a low are also being connected to the common leg at the same time. This is a no-no.

When each keypad switch is isolated with a diode, only a high signal can be transmitted from the 4017 to the common-leg line. If you don't include these steering diodes, the circuit will definitely set an Olympic glitchiness record, and you'll probably trash the 4017 the first time your finger slips on the keyboard.

The last piece of business to look at on the 4017 is why C1 and R1 are sitting on the reset input, pin 15. When this circuit (or any other one you might build) is first powered up, you want to know that things are going to start out at a known state. These two components generate a positive-going RC pulse when the circuit is turned on. That pulse, while extremely short, is enough to reset the 4017 so that the first output that gets turned on is at pin 3, the first output on the chip.

The 4518 (IC3A)—remember, we're only using half the chip—is used pretty much as it comes from the factory. You can see that there aren't many components hanging off the chip. The clock input is being driven by the same 555 output pulses as the 4017 so it counts at the same rate as the 4017. Now you might notice that we're not making as big a deal about resetting the 4518 as we did with the 4017. The way the circuit is shown in the schematic, there's apparently no way to know what state the outputs of the chip are going to be in when power is first applied to the circuit. But remember that

Looking and seeing aren't the same.

Thus, before you read further, study the schematic with open eyes and an open mind.

We don't have to build anything special to reset the 4518 because reset happens every time the 4017 cycles through its full count. The reset input of the 4518 is controlled by the first output (pin 3) of the 4017. Every time the 4017 puts a high on that output, the 4518 is hit with a high on its reset pin—and that, as we know, is exactly how you reset the 4518.

The key to making this part of the circuit work correctly is to be guaranteed that both the 4017 and the 4518 are always in sync. Every time the 4017 is putting a high on a particular output, we have to be sure that the corresponding number is showing up at the output of the 4518. Because both chips are being clocked at the same rate by the same clock, the only thing we have to do is make sure there's some mechanism to constantly check that they stay in sync with each other. By having the 4017 as the master and the 4518 as the slave (never expected kinky stuff in an electronics book, huh?), the sync between the two chips is checked every 10 clock cycles. Since the 555 is running at something like 2 kHz, our circuit is being checked for sync 200 times a second. More than enough.

When we press a keyswitch, the 4017 stops counting and stays frozen until we release the key. This is nice; but in order to have the whole circuit work, we also have to freeze the count at the 4518. The way to do this is to use the enable input of the 4518. We saw earlier that making this pin low will cause the 4518 to ignore the incoming clock pulses and freeze the current count on its output pins. We have to have this control signal generated every time a keyboard switch is closed and the 4017 count is disabled. As you probably realize, this isn't a terribly difficult thing to do.

The common-leg line of the keyswitches carries exactly the signal we need, but there's a problem—the polarity is wrong. This isn't a serious problem (even Klant Zorch worked it out), but it does mean we have to add some more silicon to the circuit.

The common-leg line signal becomes active high whenever a keyswitch is closed and gets inverted by IC4A, one-sixth of a 4049 hex inverter, before it gets connected to the enable input of the 4518. This solves every one of our circuit problems because, as things now stand, every time a keyswitch is closed, the signal freezes the 4017 and 4518; and the correct data is held on the output pins of the 4518 so some other part of our circuit can process it later on.

By now you get the basic idea and understand what all the electrons have to do for our circuit to work properly. Most of the time, all of the silicon in the circuit is knocking itself out as it counts to 10 over and over at a relatively furious rate of speed; but as soon as you close one of the keyswitches, everything should come to a sudden, screeching halt.

The way we have things set up now, both the 4017 and the 4518 come to a stop when one of the keyboard switches is pressed. This should be enough to guarantee that a valid keycode will be produced at the output of the 4518; but if you find problems or you're the kind of person who likes to have things as definite as possible, there's one more thing you can do to be sure that the circuit stops dead in its tracks whenever a keyswitch is pressed.

As you can see in the schematic of Fig. 4-7, the reset input of the 555 keyboard clock, pin 4, is tied high so there are always clock pulses coming from the chip— even if a keyswitch has been pressed and is held closed. You can modify this, as shown by the dotted line in the schematic, to make the clock stop whenever a keyswitch has been closed on the keypad.

Using the output of IC4A to control the 555's reset input means the keyboard clock will be disabled whenever a keyswitch is pressed and will stay disabled as long as the switch is kept closed. It really shouldn't be necessary to go to this extreme to have the circuit work correctly; but if you're into absolutes, this is exactly the kind of thing you'll probably want to do.

One thing you should have picked up on as we went through the discussion of our circuit so far is that the common-leg line that comes from the keyswitches is an important one. Knowing when a key has been pressed is something that we'll probably have a use for later in the design. This is more than a philosophical point because it means we have to add stuff to the circuit to prepare it for a possible journey to the uncharted regions of the future circuitry we're going to design.

There are two things we have to do to the "any key pressed" line before we can send it out of this part of the circuit. We have to isolate it from the circuit we've designed so far; and, because we're dealing with mechanical switches, we have to debounce it as well to be sure that pressing a switch once only generates one clean pulse on the line.

This may sound like a lot of stuff to do; but as you can tell by a simple glance at the schematic, it takes longer to describe what we have to do than to actually do it. Isolating the signal is accomplished by simply adding R5 to the circuit; and all we need to debounce the switches is to add C4 and R6. There are more elaborate debouncing schemes, but they're not really needed here.

If, when you get the circuit built, you find that you're still plagued with multiple signals showing up on the "any key pressed" line, try increasing the value of C4 or replacing the switches with ones of better quality.

Oops!

There's no reason in the world why you should have a problem with the circuit. Because it's so simple (at least so far), there's not much that can go wrong with it. Still, I really shouldn't say that because we all know that there aren't any certainties when it comes to breadboarding circuits.

The first thing you should do when you've finally placed all the components on the breadboard and have everything connected with wire is check for a major short circuit. This is a good habit to get into no matter what kind of circuit you're building. Because it only takes a second to do and can prevent all sorts of world-class hassles, it's rather foolish (an understatement) to overlook it.

Set your multimeter on the lowest resistance scale it has and **WITH THE POWER DISCONNECTED FROM THE BREADBOARD**, measure the resistance from power to ground. The actual number you get is unimportant—what is important is that you get some reading at all. If your multimeter reads all zeros (or down in the tenths or hundredths of an ohm), you have a problem on the board that you have to solve before going any further.

I can tell you from personal experience that shorts from power to ground are the result of a wiring screwup and have nothing whatever to do with bad components. The only way to find the source of the problem is by doing a careful eyeball examination of the breadboard. If you used the bus strips as shown in Fig. 4-9, it shouldn't take long to locate your mistake. Leaving the multimeter connected to power and ground, start lifting the wires off the power bus, one after the other, until the meter indicates that the short is gone.

Once you get a reasonable reading (something over 50 ohms or so), start reconnecting the wires, one at a time, keeping a watchful eye on the meter. Remember that, although it may be hard for your ego to swallow, it just might be true that more than one of the leads lifted from the board was causing a short. As soon as all the wires are back where they belong and there's no evidence of any short, you're ready to apply power to the circuit and see what other hassles may be in store for you.

Although the circuit is still pretty simple, there are enough wires and components on the breadboard to provide opportunities for mental and manual slipups. You can use a logic probe to test the outputs of both the 4017 and the 4518, but the best way to get an overall view of whether or not the circuit is working correctly is to hang four LEDs on the outputs of the 4518 and connect them, through 1-kΩ resistors, to ground.

THE TOP STRIP SHOULD NORMALLY CARRY POWER

THE BOTTOM STRIP SHOULD NORMALLY CARRY GROUND

4-9 Bus strips should carry only power and ground.

The outputs are counting up at about a 2-kHz rate, so unless you're a mutant there's no way you're going to see any of them blinking on and off. They should all appear to be lit; but as soon as you press a keyswitch, the LEDs should stop strobing and should show the binary equivalent of the key you pressed. But *should* is not a positive word.

If you close a switch and the only thing that happens is that you burn up a few calories from finger pressure and frustration, you've got a problem on the board. Assuming you've already checked all your connections against the schematic and made sure that everything has been stuck in the right holes (no chips are in upside down, diodes in backwards, etc.), the best way to locate the source of your problem is to start at the beginning of the circuit.

Troubleshooting stuff like this is best done by following what can be referred to as the "Wham-bam detection method." The basic rule is that you start at the beginning and work toward the end. That means your first job is to make sure the 555 is putting out a healthy bunch of clock pulses. If you've got an oscilloscope, you should see pulses that look somewhat like the ones I've shown in Fig. 4-10, a scope picture that came right off my breadboard.

If you don't have a scope, you can check for an output frequency by using a small speaker connected to ground on one end and the output of the 555, pin 3, on the other. Put a 10-μF capacitor between the 555's output and the speaker because it's not good to overload the 555's output by asking it to drive a light 8-Ω load. If you don't see anything on the scope or hear anything from the speaker, disconnect the 555's output from the rest of the circuit and try it that way. If it still seems dead, chances are you've miswired the 555 or, less likely, the chip isn't working.

The rest of the circuit can be checked the same way. If you know that the clock is working but nothing is happening at the output of the 4518, your next step is to check out the 4017. Assuming the clock signal is showing up at the 4017's clock

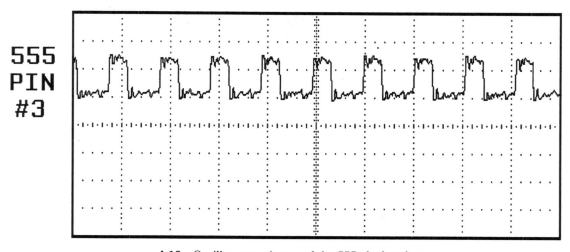

4-10   Oscilloscope picture of the 555 clock pulses.

input, the only two likely reasons for a failure here are either that the chip's reset pin is being held high or the enable pin is being held high. Either condition will keep the chip from operating.

Put your logic probe on the "any key pressed" line (the common leg of the switches) and watch what happens when a switch is closed. The line should be low when the switch is open and high when it's closed. Not only that, but as long as you hold the switch closed, you should see that the corresponding 4017 output is high and all the other ones are low. If none of this is happening, check that you've got R1 and C1 wired correctly. The last thing to check—and I do mean the last thing—is the chip itself. You can try replacing it, but you're flying in the face of the old rule that says

People make more mistakes than electrons.

Nevertheless, it's true that sometimes, occasionally, rarely, once in a while, and every so often, you run across a dead chip right out of the box from the supplier. You are using new parts, aren't you?

If everything checks out correctly to this point, the most probable source of the problem is the hex inverter. Remember that this is a CMOS part, and it doesn't take kindly to having its inputs left floating in an electronic void. The inputs all have to be tied to something; and even though I haven't shown that on the schematic, don't think you can get by without doing it.

If you don't always see signals of opposite polarity at the enable pins of the 4017 and 4518, and you're sure that all your wiring is correct, check out the inputs of the 4049 and make sure none of them is floating. Also, double-check that you've connected power and ground to the right pins of the 4049. For some reason, the location of the 4049 power pin (pin 1) is nonstandard; and a misconnection is an easy mistake to make.

So there you have it. The first part of the circuit is complete and, with a possible bit of brain damage, actually working. You can drive a golden spike through the first few boxes on our block diagram and, after a bit of R and R, get ready for the next part of the job we set for ourselves.

It's hard to imagine, but we'll have even more fun—I swear.

A slick trick

All TTL parts and a lot of CMOS parts want to be powered with a 5-V supply. Several of the CMOS parts used for the circuits in this book are like that; and when the databooks call for 5 V, they mean exactly that in no uncertain terms. If the voltage should accidentally go more than 0.5 V above the 5 V, the IC will more than likely be turned to toast.

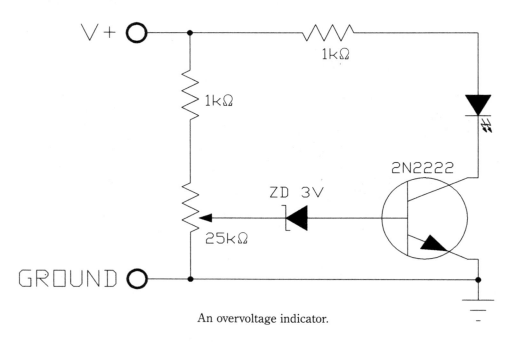

An overvoltage indicator.

Even super-regulated, bench-type power supplies can screw up occasionally and spike for a second or so. That tiny overvoltage condition can destroy a bunch of hard-to-get parts. If you add this overvoltage monitor to your circuits, you can set up the potentiometer to trigger the transistor when the voltage rises above any level you want. With the components shown, it will handle a maximum of 12 V.

5
Keyboard design #2
Bus, display, & other fun stuff

From here on out, we're on a roll. We've already talked about the kind of stuff you should keep in mind as you work through the details of the circuit you want to make, and you know what you have to have around you in terms of reference material, tools, parts, and the odd bit of nourishment (Twinkies are designer fuel. It has something to do with enzymes, I think.)

In any event, by now we've all had plenty of time to wipe the sweat off our collective brows and take care of various other matters so it's showtime at the bench once again. Things are going to move a bit faster now. We've spent enough time (maybe even more than enough time) on design basics and what you have to know to practice the fine art of creative circuit design.

We're finished with keyboard entry, and you should now have a circuit in front of you that takes any one of 10 keypresses and translates it into 4-bit binary. Just to be sure we're all starting at the same place, take a few minutes to make sure you're ready to start adding more pieces to the circuit. It's one thing to have the keyboard encoder working, but it's quite another to be sure you have a complete set of notes covering what you've done and how you've done it.

Now in the best of all possible worlds, you might think that these chapters are a complete travelogue that covers every single detail of the construction of this project. That's true, but only to the extent that I'm telling you what to do and what you should wind up with when you're done.

What I can't tell you is stuff like what color wire to use for the data outputs, power, and so on. I don't have any idea about any part substitutions, circuit modifications, or construction changes that you might have made—and I certainly don't know about any of the mistakes and unexpected hassles you had to deal with. Hey, I've got a list of my own.

No matter how detailed the instructions for a project you're trying to build, and no matter how closely you follow those instructions, every project becomes the personal property of the person doing the building. Put two people in the same room doing the same thing and you'll wind up with two different end products.

As you work your way through these pages, you should be producing a design notebook of your own. Keep a record of the mistakes you make, the ideas you come up with, and any variations you think you might want to try later on. Notes like these should be kept for every project you do and filed carefully away when you're finished with the job. Let me tell you, the more work you do, the more you'll find that this library of circuit descriptions, ideas, and pitfalls is the most valuable set of resource materials you have. And you can't buy them anywhere (except on Neptune).

Controlling the digits

The keyboard encoder we finished is a neat circuit, but even Cro-Magnon people living peacefully in an undiscovered valley on Bora Bora would realize before long that the circuit falls pretty far short of being really useful. It just handles one digit at a time; and because the output bus is only 4 bits wide, as soon as you enter a second digit the first one vanishes into hyperspace.

Since the design criteria specify that the final circuit has to be able to handle up to 10 digits, it's clear that things are far from complete. When you're talking about 10 different entries from the keyboard, the first thing that has to be done is to work out some way to distinguish between them. We already recognized the need for something like this. As you can see in Fig. 5-1 (the updated block diagram), this is the section we labeled the Digit Selector. We've already crossed off three of the circuit sections shown in the block diagram. However, if you take a good look at the diagram, you'll see that before we can start any work on the rest of the circuit we've got to focus our attention on the digit selector. All the remaining elements can only be designed when we have some way of telling the circuit that this is keypress number 1, this is keypress number 2, and so on, up to and including keypress number 10.

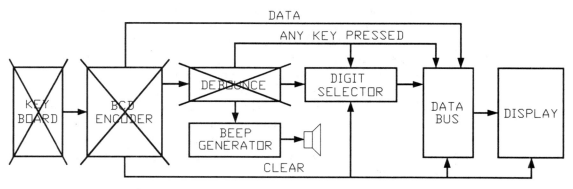

5-1   Block diagram of the keyboard with the digit selector.

Starting off a fresh part of the circuit means first thinking about what it has to do and what has to be done to make it do it. Because the digit selector has to keep track of how many times we press a key on the keyboard, it's evident that we need some sort of counter. Besides counting the keystrokes, the circuit has to give us an unambiguous, unique signal for each count.

By this time it should be clear that we already know about a chip that's an ideal choice for the job—of course, I'm talking about the 4017, the same chip we used as a keyswitch decoder. It takes an input clock and gives us up to 10 mutually exclusive outputs that can be used to identify exactly what keypress we're up to.

It would seem that there's nothing special to keep in mind when we lay out this part of the circuit. After all, all we have to do is feed the 4017 with a signal that tells us when a key has been pressed and let the chip do its thing. In actuality, the more time you put in at the bench, the more you learn that not only is nothing ever that straightforward, but there are times when things get unbelievably sneaky as well. There's some sort of rule about this—something eternal . . . like the Law of Gravity.

The circuit shown in Fig. 5-2 should be familiar to you. The 4017 digit selector is being clocked by the "any key pressed" line we set up on the keyboard and, as stated on page 12 of the manual, should give us a unique output for each keypress. This is all well and good, but if you take a close look at the circuit and recall what we were going over in the discussion we had about numbering outputs, you'll notice something strange about the sequence of the output pins. The output aimed at the first digit is coming from pin 2, which also happens to be output number 2 on

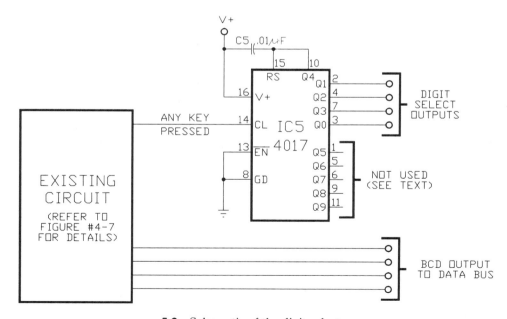

5-2 Schematic of the digit selector.

the chip. Not only that, but output 1 is indicated as being the identifier for the count of the last digit entered from the keyboard. Pretty strange stuff.

Now it may seem weird but, as we all know, there's a logical explanation for everything—except this (just kidding).

The answer begins with C5, a small capacitor that provides a reset pulse to the chip when the power is first turned on. We need something like this because the 4017 has the annoying characteristic of usually assuming a stupid state when it's first turned on. Try it for yourself and you'll see what I mean—several outputs will be on at the same time, no outputs will be on—the only thing you'll be able to count on is that you won't be able to count on it. Fortunately, that's a simple problem to solve.

At power-up, C5 resets the chip and, as we all know, that means the first output, pin 3, is made high. Remember that this happens at power-up, before any entry has been made at the keyboard. No signal has been sent from the keyboard along the "any key pressed" line. When a key is pressed and legitimate data gets put on the bus shown in Fig. 5-2, a positive-going pulse is also put on the "any key pressed" line. This causes the 4017 digit selector (IC5) to select its next output—in this case, on pin 2.

The bottom line here is that once IC5 has gotten a power-up reset, the first keypress at the keyboard will result in a high on the second output of the digit selector. This is only a big problem if you don't know what's going on. Take it from me, until you figure out what's happening and why it's happening, you can lose a lot of hair.

Chalk this kind of stuff up to the old saying

Nothing does what you expect it to do.

What it does do is just different enough to guarantee you a three-week vacation at the funny farm, where you can split your time between finger painting and getting your head candled.

As things stand now, we have signals to indicate the sequence of keypresses, but we're still short of having something that can be used in the circuit. The best way to understand what we need is to start thinking about the 4-bit data output bus.

Bus and display

We specified that we wanted to be able to see the digits we entered on an LED display; thus, we need something to drive the seven-segment displays we'll be using. Because we also specified that all the parts for the circuit would be cheap and available, we should be using a standard display driver IC that can be gotten anywhere. One good choice that meets every one of the requirements is the 4511, shown in Fig. 5-3.

Let's go through a quick rundown of the pins and then see how we can use the chip.

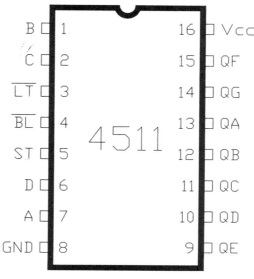

5-3 Pinouts of the 4511.

Output pins   The output pins are pins 9 through 15. (In keeping with the usual confusion, the order of the pins isn't the same as the order of the outputs.) The 4511 is designed to drive common cathode displays. Each output can only supply about 25 mA or so; thus, you have to be sure to use a resistor to limit the current through the LED segments.

Input pins   The input pins are scattered all over one side of the IC. To list them in order, they're found at pins 7, 1, 2, and 6. The inputs are expecting to see standard weighted binary data—the sort of stuff supplied by the 4518 we're using in the circuit.

Blanking input   The $\overline{blanking}$ input is on pin 4. It's an active low control input, and putting a low here will cause all the outputs to go low and blank the display. In most operations this pin is tied high, but it can also be tied to external logic and used to blank leading zeros from a multidigit display. One neat use for this input is to feed it with the output of a variable-duty-cycle oscillator so you can control the display brightness.

Lamp test input   The $\overline{lamp\ test}$ input is on pin 3. If you make this input high, all the segments in the LED will light regardless of the code presented to the inputs or what you're doing with the blanking input. If you put a manual control on this pin, you'll have the ability to press a button and make sure that all segments in the display are still working. If you design your circuit properly, you won't have to worry about stuff like that.

Store input   The store input on pin 5 is the control for the 4511's internal latch. If you put a low on this pin, the latch will be open and the outputs will follow the inputs. Putting a high on the pin closes the latch, the 4511 ignores the inputs, and the outputs remain at whatever state they were in the last time the store input was high.

Other pins   The other pins are the power and ground inputs at the normal pin 16 and pin 8 locations, respectively. The 4511 is a CMOS chip so it will work at any voltage from 3 to 15 V—and yes, the brightness of the display is directly related to the value of the supply voltage.

Now that we have the players in position, all we have to do is work out how to connect them. At first glance, the most obvious way is simply to connect the 4511 inputs to the corresponding outputs of the 4518, tie the store input to the output of our digit selector, and then cross off a whole bunch of boxes in the block diagram.

Not so fast. It would be great to do this (and you can do it if you want), but there's one big problem—it won't work. You should be able to see why; but in case you don't, let's see what would happen if things were connected like this.

The first problem we'd have (and this one is big enough to kill the whole idea) would be that as soon as we pressed a key, the output of the keycode encoder would freeze with the binary equivalent of the keyswitch. The pulse on the "any key pressed" line would clock the digit selector and that, in turn, would enable the store input of the 4511. A few microseconds later, the number we selected at the keyboard would immediately appear on the display.

Now for the bad part. Remember that the only time the keycode encoder has a static output is while a key is pressed. If your fingers are busy doing something else, the encoder's outputs are constantly scanning across all of the possible keycode values—from 1 to 10. Everything works while we have a finger pressing a switch on the keypad; but the moment we lift our finger, the 4518 encoder starts scanning again. Because the 4017 digit selector is keeping the 4511's latch enabled, the display will start scanning from 1 to 10 at the keyboard clock rate— somewhere about 2 kHz. An interesting effect—not terribly useful, but interesting.

The reason this is happening is obvious: the outputs of the digit selector stay high after the digit is selected. What we have to do is provide some way for the digit selector to only send an enabling pulse to the 4511 instead of holding the 4511's latch open. It's clear that the digit selector has to be isolated from the 4511. Its outputs have to be triggers for a pulse generator that will control the latch. We can't let the outputs control the latch directly.

Now pulse generators are a dime a dozen. There are lots of ways to build them, and you've more than likely put a few of them together yourself. They can be anything from a simple resistor and capacitor (such as the one we used on the reset pin of the keyswitch encoder), to a 555-based circuit, to a whole host of other stuff.

Because we're going to need a pulse generator for each of the digits we're able to enter (that's a maximum of 10), it's not a good idea to go nuts on the design. In this case, "going nuts" means either spending a lot of time on it or making the thing too involved.

The simplest choice is to use a resistor and capacitor, but these can be a problem because the charge and discharge times of the capacitor result in really sloppy rise times—both going up and coming down. A good alternative is to use an RC circuit but add something to it that will clean up the shoulders of the wave and make the waveform a nice, sharp-cornered, square wave.

We've already got an inverter on the board (IC4A), and we're only using one of the inverters in the package. That leaves five of them sitting there doing nothing else besides soaking up electricity. If you plan on running off batteries, that's what we refer to in the technical journals as a real waste. Admittedly, unused inverters only use flea power but remember:

Don't waste electrons.

Before you know it, you can wind up with a whole board full of freeloading silicon—a good thing to avoid. As luck would have it, inverters are a good choice to use when you want to build a bunch of down-and-dirty pulse generators and you need them in a hurry. This circuit setup is used so often that it even has a name—half monostable. When you use an inverter, there are two possible ways to build half monostables and both are shown in Fig. 5-4. The nice thing about these is that, as you can see, they will generate either positive- or negative-going pulses; and what's even better, there aren't too many things to watch out for when you're putting them together.

The key to these things is that CMOS parts are wonderfully immune to noise. Anything over half the supply voltage will be seen by a CMOS input as a high, and anything less will be seen as a low. When the capacitor charges, it only has to get halfway to the power or ground rail to make the inverter snap its output.

You'll notice that the output pulses are inverted. This makes sense because we're building around an inverter. You can make a half monostable out of noninverting buffers (like the 4050) and have the output pulse be the same polarity as the input pulse. The only care you have to take is to make sure the input pulse is longer than the length you want for the output pulse. If you think about it for a bit, you'll realize that this is common sense because the inverter is just working to

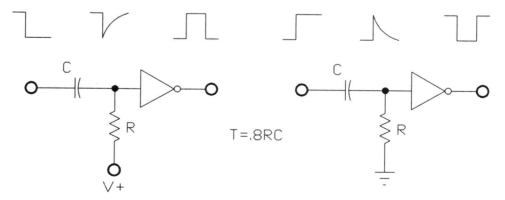

5-4   Generating positive and negative pulses with inverter-based half monostables.

clean up the charge/discharge curve of the capacitor, and the output of the inverter will change state when the capacitor charges or discharges.

In any event, our 4017 digit selector is the perfect choice for an input to a half monostable. When its outputs are active, they go high and stay high as long as they're selected. The period of all the output pulses, regardless of whether you build the half monostables with inverters or not, is roughly equal to 0.8(RC).

But back to the problem at hand. In order for us to solve the problem we were talking about earlier, we have to put properly designed pulse generators between the outputs of the 4017 digit selector and the inputs of the latch controls (the store inputs) of the 4511s. This may strike you as a real pain in the neck, but it's just part of the always wonderful world of electronic design. The more time you spend in this business, the more you learn that

Nothing is hassle-free—ever.

While the hassles may be big ones, small ones, or just plain old Goldilocks ones, they're always waiting to pop up when you least expect them.

Putting it together

This is the perfect place to use a sports analogy. You know what I'm talking about—something like: Now that all the players are on the field . . .. I stopped doing that kind of stuff after I read a literary review of a book about Charles Dickens by some guy (a name I've luckily managed to forget) who started his review by saying "I just went three rounds with Charles Dickens and he was KO'd in the third." I know this doesn't have anything to do with the keyboard circuit, but it was on my mind. Sorry about that.

Every piece we need for adding a display and building a bus has been designed and gone over in detail. The time has come to put everything together and see what we forgot to do. You might have expected me to say something else, but remember that there's always a possibility of finding some sort of circuit insanity lurking around the next corner. That's the right attitude to have when you're in the breadboard stage because

No expectations means no surprises.

Nobody wants to call in the admiring public until he or she is really, absolutely, positively, it's-already-worked-60-gazillion-times, sure of the circuit.

The circuit we're going to put together is shown in Fig. 5-5, and the placement diagram for all of you using solderless breadboards is in Fig. 5-6. You'll notice that I've listed FND500s as the seven-segment display. You can use any ones you want or happen to have on hand (as long as they're common cathode displays). I like to use the FND500s whenever I'm working on a breadboard because the LED pins go across the top and bottom of the display (Fig. 5-7). The displays sit on the bread-

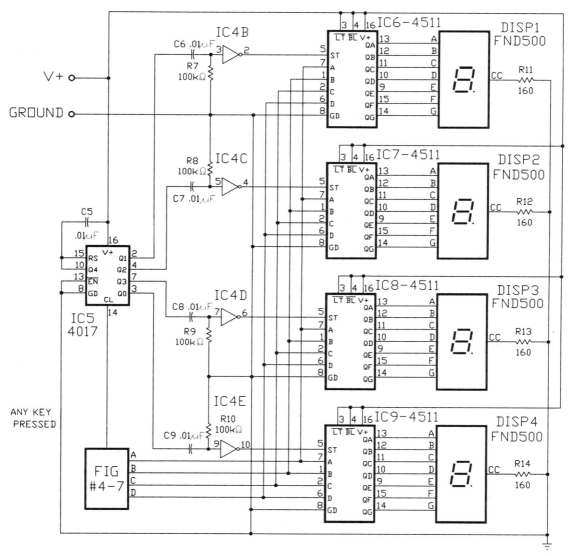

5-5 Schematic of the circuit with the display.

boards, the digits facing the right direction; and each pin is on a separate five-point data bus. The LEDs aren't quite as bright as some others, and (from what I've seen) red has been falling out of vogue (yellow seems to be in); but anything that makes life easier on the bench is something I'm in favor of. Besides, I have lots of them in my drawer.

The only things in the schematic of Fig. 5-5 that we haven't had a reason to go heavily into are resistors R11 to R14. These are the current limiters for the displays; and while the common practice is to put a resistor on the line between each

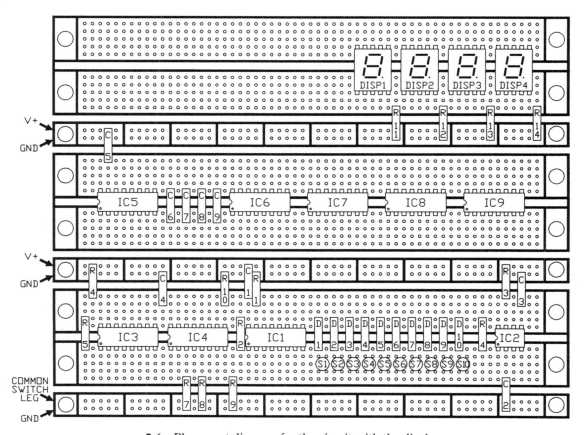

5-6 Placement diagram for the circuit with the display.

4511 output and its related display input, you can do it this way as well. Because the purpose of the resistor is to limit the current, it's perfectly acceptable to put a single one on the cathode. It makes the breadboard simpler (always a good thing) and uses fewer resistors. Before you make this part of your electronic habit, however, understand that you'll pay a price for this bit of circuit simplification.

The reason for using separate resistors on each segment (seven per display) is that equal amounts of current flow through each segment. That means the display brightness will be the same, no matter how many segments are being used to display a particular number. When you put a single resistor on the cathode, you may still be protecting the display but you're putting a limit on the current that can flow through the entire display. If two segments are lit, they'll each use half that current; if four segments are lit, each segment will use a quarter of the current, and so on.

The brightness of the display, therefore, will depend on the number that's being displayed. Numbers that need fewer lit segments will be brighter than numbers that light more segments. This can make for a very annoying display, particularly if numbers change rapidly, but it's much easier to do things like this when the

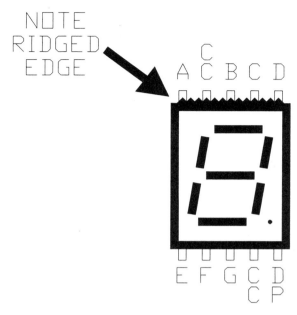

5-7 Pinouts of the FND500.

circuit still lives on the breadboard. Evening out the display brightness is the kind of stuff you worry about when you've put a lock on the circuit and are laying it out in its final form—on a PC board or something like that.

The last comment to make about the schematic is to apologize for cutting it short. I've only put four digits on the board because more would make the drawing much too crowded and wouldn't give you any more information anyway.

The only difference on the board between a circuit to deal with 10 digits and the one shown that deals with four digits is what happens with IC5, the digit selector. The schematic of Fig. 5-5 shows the reset input, pin 15, connected to output 5 (pin 10). When things are connected like this, the digit selector will reset after the third digit has been entered from the keyboard and immediately put a high on the first output at pin 3.

It's important for you to understand the counting sequence of the digit selector as shown in the schematic. When the chip is first powered up, the reset pulse generated by C5 causes it to put a high on pin 3, the first output. This is a problem because it happens even though no key has been pressed on the keypad. When the first keypress is done at the keyboard, the "any key pressed" line signals the digit selector and IC5 clocks over to the next output (output 2 on pin 2). This may be the first digit entered from the keyboard, but the digit selector signals that on output 2.

You'll remember that the reason for this problem is that there's no way to disable all the outputs on a 4017. There will always be a high on one output or another.

As digits are entered from the keyboard, the digit selector will be watching the "any key pressed" line and will sequence to the next output in line. If, as in the case of the schematic, you only want four digits in the circuit, stop and consider what happens after three of them have been entered.

Just before the fourth digit comes down the line from the keyboard, the 4017 digit selector has its fourth output (pin 7) high. When the fourth digit is entered, IC5 is clocked, advances its count, and puts a high on its fifth output (pin 10). Since we have this output connected to the reset pin, the high causes the chip to immediately reset and put a high on pin 3, the IC's first output. This, as you can see, causes the latch associated with display 4 to be addressed and everything works correctly.

If you're not sure about what's happening, take the time to go over this until you understand it completely. If it all seems confusing to you, temporarily connect the digit selector in what would seem to be the logical order. This would be the first output of IC5 to the 4511 driving the first digit, the second output to the second digit, and so on. Leave the reset pin connected to the fifth output at pin 10 and watch the behavior of the circuit. Everything will become clear to you.

If you want to use 10 digits on the breadboard (although all it's going to do is confuse things), the only change you have to make is to connect the reset pin to ground. The 4017 will automatically reset as soon as the number of entered digits passes 10.

Oops!

Once you add this stuff to the breadboarded keyboard encoder from the last chapter, you should have a much more useful circuit. Of course, the only way you can tell is if everything works correctly, and we already know that's just not the way things usually work on the planet Earth—or anywhere else for that matter, Neptune included.

Assuming that you made sure the earlier circuit was working before moving on, just about the only thing that can be wrong with this one is something mechanical. The only electronic thing to watch out for is that there should still be one unused inverter on the board. As we did earlier, the input has to be tied either high or low. If you leave it floating, you're not only asking for trouble but you're more than likely to get it as well.

Everything we've added to the breadboard can easily be checked with a logic probe because the additions to the circuit are all static—no lines will move unless a trigger pulse is sent down the line from the keyboard. As a matter of fact, if you do see movement on any of the lines (with the exception of the four-line data bus), you've got something miswired.

The data bus we've made should be constantly moving and should stop dead in its tracks as soon as you close any of the switches on the keypad. The best way to run down a problem is to keep a key pressed on the keyboard and make sure that the only line changing state is the output of the keyboard clock. If you elected to freeze the clock whenever a key is pressed (see the last chapter), pressing a key will kill everything and make the entire circuit static.

If you own some exotic and rather pricey test equipment, you can examine the outputs of the half monostables and make sure a pulse is showing up there when a key is pressed. However, there are simpler tests you can make using much simpler equipment. Take a look at the outputs of the digit selector and see if the outputs sequence as you press keys on the keypad. If nothing is happening there, the next place to look is the "any key pressed" line. You should see it change state when a key is pressed.

A lack of activity on the "any key pressed" line means you've got a problem with the circuit from the last chapter. You should have taken care of that much earlier, but it's always possible to pull a wire loose when things are still in the breadboard stage. I usually use a pair of tweezers when I'm adding connections to the board because fingers make it hard to see exactly which holes I'm putting the wires into. It's easy to make the kinds of mistakes that can take hours to find and correct—lots and lots of hours.

In general, you should use the same wham-bam detection method for troubleshooting that we used in the last chapter. The key points to look at are the outputs of the digit selector, the store inputs of the 4511s, and the four lines on the data bus. Check to be sure you have good continuity where you should and none where you shouldn't. All the inputs of the 4511s should be paralleled on the data bus; and if they're not, you've got the possibility of a floating input.

If you suspect any of the displays, ground each of the lamp test inputs in turn. If all the segments of the associated display light up, everything's OK; but if they don't, you've probably made a bad connection.

The last thing to try is replacing the ICs. As we all know by now, the most likely source of a problem is a screwup on your part. Swapping chips is usually nothing more than an act of desperation on the part of the designer.

So now what?

I'm sorry to tell you this but (now that you've got the circuit under control and can see what it does)—and you've probably already seen it for yourself—the circuit leaves a lot to be desired. We're a lot closer to the goals we set for ourselves when we started this design exercise, but a glance at the block diagram will show that there's still a bunch of stuff left to do.

For openers, there's no way to reset the display and the number you see on the display at power-up depends completely on the particular 4511s you're using. For all intents and purposes the initial numbers appear at random, and the chances of getting a zero or a blank are up to the great god Electro.

We need a way to clear the display from the keyboard. The only option we have is to enter a bunch of zeros from the keypad—not a method that you would exactly call elegant. We're also missing a real data bus. The only place we have the four digits saved as individual data is at the LED displays, and decoding the seven-segment information isn't even something they would do on Neptune. A better way is to have the data collect on a bus and let the LEDs be set up as peripheral circuitry whose function is only to display the data, not store it.

But everything we've done here is needed before we can add any more stuff to the breadboard. So get it all working and, once you do, add a few more check marks to the block diagram, turn off the power, and go out to see a mindless movie.

There's more left to do but you'll love it.

A slick trick

I've been using this design for years whenever I have a circuit that has a battery to back up data after the main power is turned off. The data to be backed up can be anything that needs to be saved—memory data, counter states, or logic conditions in general. No matter how many times I have this little circuit published, I keep on getting letters that ask me for it.

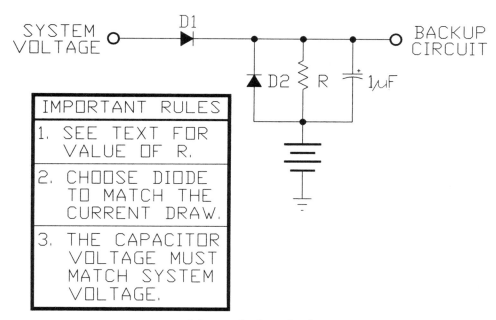

A battery backup circuit.

The value for the current-limiting resistor is different for different batteries but can easily be calculated for your battery by simply applying Ohm's Law as I've shown.

$$R = (Vdd - 0.6 - Vbat)/Ibat$$

The 0.6 figure is the voltage drop across D1, and Ibat is the current you want to use for charging the battery. A standard charging current value for a battery being used for memory backups is one-tenth the amp-hour capacity of the battery.

6
Keyboard design #3
Beeping, busing, & brainstorming

There are always favorites. No matter what kind of stuff you're talking about, everybody's got heroes and villains. Now that's as true (probably truer) for me as it is for anyone; and since I'm an eclectic kind of guy, my list of heroes and villains is really long.

Now I don't know how many of you people out there are into history, but I've always liked reading that kind of stuff. Maybe it's because I get some kind of personal satisfaction from finding out that there were people who, when faced with problems bigger than mine, screwed up more than I would. It's amazing how stupid some people can be.

One of the greatest pieces of historical (?) writing I ever read was Homer's *Odyssey*. Talk about problems, here was this guy Ulysses who lost a war, lost his way, and almost lost his kingdom. Not only that, but he had to deal with gods who, for some reason or another, took a personal interest in making everything as tough for him as they possibly could. Gods were different back then, I guess.

Despite all the swashbuckling stuff in the story and a cast of guys like Hercules, my favorite character is Ulysses' wife Penelope.

Now catch this. Ulysses is away for some 20 years and a load of what Homer referred to as suitors (but today would be called freeloaders) are all hanging around Ulysses' wife trying to convince her that Ulysses is dead. If they can do that, one of them can marry her and become king. Anyway, Penelope (who has to be the ultimate military wife) tells them that she'll start weaving this tapestry; and if Ulysses doesn't show up by the time she's finished, she'll marry one of these guys.

So there you are. The suitors spend every day for more than 20 years stuffing their faces (these guys are eating everything including the dishes and probably—Homer isn't clear on this—the paint off the walls) and watching Penelope weave this tapestry. The reason my hero (heroine?) is Penelope is that she would weave during the day while the suitors were watching, and—get this—unweave at night.

I admire Penelope—her devotion to her husband, her confidence in his ability, and her faith in his return—but the reason she became my hero is that she pulled off what has got to be the biggest scam of all time. I don't care how stupid her suitors were—and maybe they were eating the paint—but because they never figured out what Penelope was doing, she had to be incredibly slick.

I know I'm stretching this but, believe it or not, there really is a sort of relevance between this story and the work we're doing on the keyboard encoder. Sorta, kinda.

One of the reasons for drawing up a list of design criteria when you start a project is to help define project goals. Once you know where you're trying to go, you not only can get there a lot faster but you'll have no trouble knowing when you've finally arrived. Even if you're checking things off a list, there's a real tendency not to stop when everything on the list has been completed.

There's nothing wrong with adding to or changing goals during the course of circuit development; but if you're not careful, the project will take on a life of its own, and you'll never reach the end of it. We set out to build a generic keyboard circuit. As we work our way through the design, I'm sure lots of thoughts cross your mind about what could be added or modified; but you're better off jotting those thoughts down on paper and holding off any work on them until you have the original circuit working on the bench.

I know that the stuff you thought up is world class and that it's going to change the world, but hold off doing it until we get through this chapter.

I know this is your book and you're the one making the investment in time and money to go through it, but let me tell you that if you go overboard with changes and stuff before you actually finish the basic design, you're running the risk of not finishing at all. The biggest danger on the bench is ego, not electrons. The majority of products never get completed because the designer loses sight of one basic rule that says

First get it working, then get it good.

You can get so sidetracked by changes that the project will be obsolete before you finish it—if you ever do.

Beeping

Since we're on the subject of sticking to design criteria, you might remember that one criterion included in the original list was to have a beep sound whenever a key was pressed. This feature is a nice thing to have in a keyboard circuit because it provides a definite indication that a key was pressed. We've already crossed a bunch of boxes off on the block diagram (Fig. 6-1), but the beep generator and a few other things are still left to do.

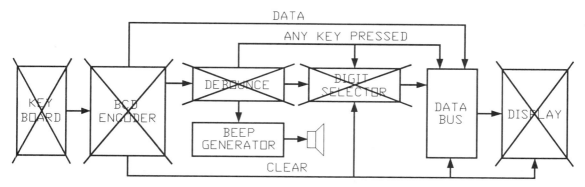

6-1 Block diagram of the keyboard.

Even though an audible keypress signal is a noticeable feature of the circuit, it's not one that should be done if it's going to take a lot of time and silicon to create. We're free to take as much time as we want on any part of this project; but when you get out in the real world, you'll find that one of the major ways to measure results is in hours of development time and circuit complexity.

There's no getting around the fact that some things take a long time to develop. Complexity is frequently unavoidable. But when you start work on some parts of a circuit, you always get a basic feeling in your gut that tells you how complicated things are going to turn out to be and how long it's going to take to get them done.

The job in front of us is to add a beep to the circuit. While gut feelings are the result of experience, even a rank beginner should be able to recognize that this is a fairly simple thing to do without adding too many parts to the circuit. Any circuit designer will tell you to

Listen to your gut.

However biologically backward that may be, in this case it's exactly the right feeling to have.

When we first started this design and were working out the details of the keyboard clock, we picked 2 kHz as a reasonable frequency. It was faster than even the Olympic gold medalist in typing and was nice and easy to set up. Now that we're talking about adding a keyboard beep, it turns out that 2 kHz is a good choice for the beep frequency as well. Since we now have a usable frequency available for us in the circuit, a large part of the job is done. The only thing left is to figure out how we want to make it happen.

For openers, we know that the beep should sound whenever a key is pressed on the keyboard. This means that the trigger for generating the beep has to come off the "any key pressed" line. The design job in front of us is to lay out some cir-

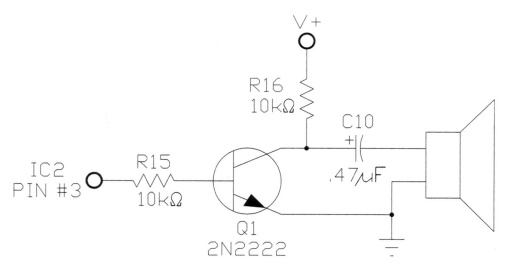

6-2 Basic transistor-based beep generator.

cuit, keeping it as simple as possible, that will transfer the keyboard clock to a speaker whenever it's told to do so by the keyboard signal.

Figure 6-2 provides one answer to the problem. The transistor is used as a switch to turn on the speaker. Notice that it isn't set up as an amplifier. You could do that if you wanted to, but there's no need to because the output of the 555 has enough juice to drive a speaker directly. All we have to do is use the transistor to turn the speaker on and off.

As shown in the schematic, the output frequency from the keyboard clock is isolated by R15 and fed to the base of the transistor. You might be able to get by without adding something like R15, but it's always a good idea to electronically separate different sections of the circuit that are being triggered by the same signal. When you're still in the developmental stage, it's better to make it a rule to always take the most conservative approach to design.

If you're using one signal to trigger several different parts of a circuit, keeping them isolated by resistors is an attempt to keep the operation of one circuit section from interfering with the operation of another. It's not always necessary, but it never hurts.

The collector of the transistor is sent to the speaker through C10 and, in most cases, all we'd be missing to make the whole thing work is some voltage on the collector.

Since we want the beep to sound when a key is pressed, we can take our beep trigger from the "any key pressed" line. We're still missing a voltage source for the transistor's collector so that's just as good a place as any to connect the trigger that turns on the beep. We'll be needing a positive signal there; and while the "any key pressed" line is active low, we can take the signal from the output of IC4A, the inverter that inverts the "any key pressed" line to control the enable input of the 4518.

Now I have to tell you that what we're doing isn't strictly kosher because, in the normal course of things, asking a CMOS output for enough juice to power a transistor—even a small signal one—is not exactly the kind of thing you'll find as a recommended procedure in the databooks. We can get away with it here because the 4049 has an unusual amount of output drive capability.

So it would seem that we have everything under control. All that we have to do is tie the inverted version of the "any key pressed" line to the collector of Q1 and call the design a wrap.

You can wire it up this way if you like; but when you do, you'll find that while the speaker will sound a 2-kHz tone when you press a key on the keypad, the tone will stay around for as long as you keep the key pressed. Not too swift.

A better design would be one that sounds a beep of predetermined length for each keypress. There are several ways to make this happen, but it would be nice to hunt around the circuit and see if there's any spare silicon available. Remember that we wanted to keep from adding any unnecessary components to the board. The only thing worse than an overpopulated board is one with lots of unused gates and other bits of miscellaneous silicon. Both conditions indicate a poor design.

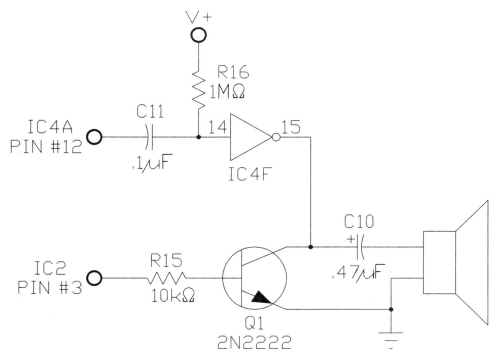

6-3 Final beep generator circuit.

We do happen to have a spare inverter available. If you go back through the last chapter, you'll see that we spent some time getting acquainted with half monostables. They take an input trigger and send out a pulse of easily configurable length. Since that's exactly what we want to do with the beep generator, all we have to do is set the last inverter up that way and choose components that will give us a beep duration that seems adequate. The values shown in Fig. 6-3 produce a pulse of about a tenth of a second. That's long enough to be noticeable without being offensive.

If you want to make the pulse longer or shorter, this is the time. The change is such a minor thing, you'll probably forget it if you leave it till later.

You can put this circuit on the breadboard as shown in Fig. 6-4 even before you work on the rest of the circuit we'll be developing later in this chapter. You'll have a good indication of a keypress, and the data you enter will show up on the display you've already wired on the breadboard.

If you're having any trouble with the circuit, a beep for each keypress is a simple way of helping you find out which part of the circuit is working and which part

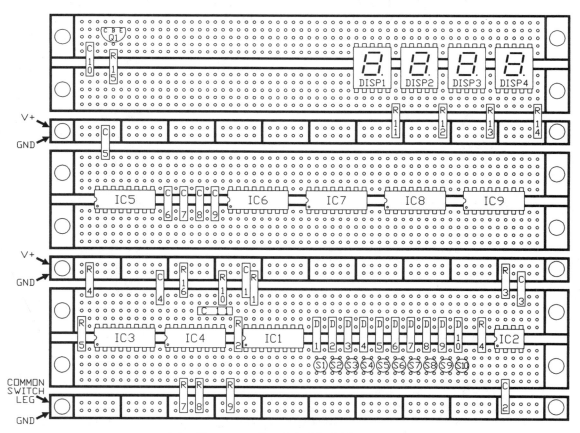

6-4 Placement diagram with the beep generator.

is in trouble. Remember that the beep generator is really a pulse detector—a useful piece of debugging gear. If you want to use it to check whether or not pulses are showing up in other parts of the circuit, you can either build a second one on a separate breadboard or use the one you have by moving the connection you made to IC4A pin 12 to whatever other part of the circuit you want to test.

Keep an eye on the polarity of the pulses you want to test. The beep generator we made for the keyboard is designed to detect negative-going pulses. If you want to use it to detect positive-going pulses, you have to move the Vcc end of R16 to ground.

Storing data

If you play around a bit with the circuit we've put together so far (shown in the schematic of Fig. 6-5), you'll soon recognize that it's not really the kind of thing you want to call anybody in to admire. Some of the things that are missing are minor (and we'll get to them), but there's one big glaring place that still has to be developed. We put it in our list of design criteria, but this is the first time the developing circuit has been far enough along for the missing stuff to be noted.

As things stand now, we can enter numeric data from the keyboard, but it's not being stored in any manner that makes it possible for us to feed the data out to the real world. As a matter of fact, the only place our data is being stored is in the display; and while that might be convenient, it's really not useful when you want to get the data off the board. And you will want to do that. After all, that's really the purpose of this circuit.

Adding storage to the circuit starts with the simple statement that we need something to store the data in. We'll undoubtedly have to add some silicon to the board because the only storage devices we have now are the 4511's internal latches. These latches are in use, but even if they weren't, they're not suitable for the kind of storage we want to do.

Once we work out how we're going to store the keyboard data, we also have to decide how we want to get the data out to the real world. The bus configuration we're going to need depends on whether we want to build a serial or parallel bus. This is a basic decision whenever you set out to create a data bus. While you certainly can change whatever you want whenever you want to, it's always a good idea to give the matter a bit of serious thought before deciding.

Once upon a time, just about every keyboard output bus was built as a parallel one. The parallel bus not only made the design much more straightforward, but (as in the case of computer keyboards) it provided an agreed-upon standard—an 8-bit output bus from the keyboard presented the computer with ASCII character codes. Back in the days of the CP/M and Apple II, this is exactly the way things were done. The Apple IIe changed over to a nonstandard output bus because an encoder IC was put on the motherboard and the keyboard was redesigned to be a simple XY matrix of switches. Apple claimed it was a better design. It was also a cheaper design.

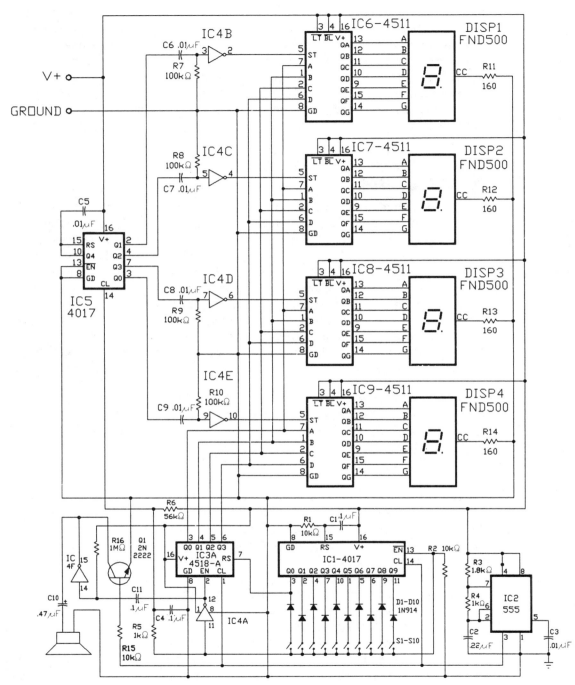

6-5 Current schematic of the circuit.

Then IBM entered the computer market and decided to do away with the idea of an ASCII keyboard. Data was sent down the wire to the computer on a serial bus as a steady stream of single bits. The reasons for this change are buried in a vault at IBM headquarters, but I'm sure they make interesting reading.

If you open up an IBM-compatible keyboard, you'll find it's really a small computer with a microprocessor (something like an Intel 8247 or 8248 controller) to read the keyboard switches and another bunch of circuitry to send the serial data to the computer.

For our purposes, the choice of a serial or parallel output isn't very important because we're not designing this keyboard for any one application in particular. This is a general design that we can stick in a notebook and use later whenever we have a need for it. And keep in mind that it's no big deal to convert serial to parallel and vice versa.

It may take a bit of extra brainwork to do the conversion, but once you have the basic keyboard design working and safely stored away on a page in your notebook, you'll feel much better. Remember that both the serial and parallel keyboard designs need all of the keyboard elements we've been talking about. All that's different is how they send their data out to the world.

But since we have to do one or the other, and since a series bus is a bit more complex to build, and since even a serial bus has to be fed by a parallel one, and since we're all familiar with the design rule that says—

When in doubt, keep it simple.

We might as well build a parallel bus for our circuit. Got it?

Parallel buses shouldn't be a new thing for you. As a matter of fact, if you look at Fig. 6-5 (the schematic of the circuit we've been building), you'll see that we've already built quite a number of them. One is the internal 4-bit bus that feeds the 4511s, and the others are the display buses that drive the LEDs. Creating an output bus is simply a matter of deciding where to get the data and then designing the circuitry to store it.

The first decision is choosing the storage device. This can be a series of latches, memory, or anything that can hold the data entered from the keypad. Each keypress we make generates a 4-bit word; thus, one of the basic requirements of our storage system is that it has to be able to store 4 bits as a single unit. Because we've also specified that we want to be able to handle up to 10 digits at a time, another requirement is that we should be able to expand the storage system so that it can handle as many digits as we decide are needed for any application using this keyboard design.

The easiest way to store the keyboard data is to have a separate latch for each digit. If we use a single memory chip to do the job, the component count would be less but the support circuitry to get the data in and out of memory would be much more complex.

You can see what I mean because ten 4-bit pieces of data makes 40 separate bits; and the only place you can get a memory chip that has a 40-bit data bus is, you guessed it, on Neptune.

I'll make you a promise—if you'll make the trip, I'll design the circuit. Just drop me a postcard when you get there.

Until Earth scientists work out the details of warp drive, I'm sad to admit that we'll have to make do with the parts of suppliers here on our own planet, however primitive their stuff might be. Once we accept these limitations, there are a lot of devices suitable for us to use. Even considering our requirement that all the parts have to be cheap and available, there are several ICs that we can use in the circuit.

All things considered, the 4508 (Fig. 6-6) is a CMOS 8-bit latch that's ideally suited for our circuit. The outputs can be tristated, and each chip is really two 4-bit latches in a single package (which makes it terrific for our application). Each half of the chip is a completely separate device; but when the control pins are connected in parallel, the 4508 becomes an 8-bit latch.

The pinouts for the 4508 are pretty much self-explanatory; but if you don't have a data sheet or book (a very serious problem), here's a brief rundown. Like the 4518, each half of the chip is exactly the same so I'll only list half the pin numbers.

Data inputs The data inputs are on pins 4, 6, 8, and 10. Even though databooks label the input pins with numbers like D2 and D3, they're all the same.

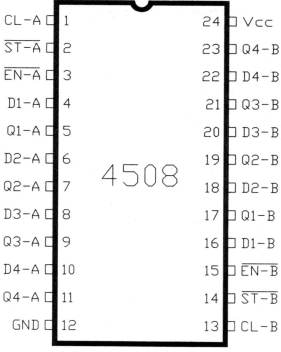

6-6 Pinouts of the 4508.

Remember that this is a storage device, and the only reason for numbering the input pins is simply to relate them to particular output pins.

Data outputs The data outputs are on pins 5, 7, 9, and 11. Each output pin is related to the input pin just below it (output pin 5 is tied to input pin 4, and so on), and all of them can be tristated.

Clear input The clear input is at pin 1. Putting a high on this pin resets all the latches to zero. During normal use this pin has to be tied to ground.

Store input The $\overline{store}$ input is on pin 2. This is the control for the latch and is active low. When this input is high, the latches will become transparent, and whatever data is fed to the inputs will show up at the outputs. Making the $\overline{store}$ input low will cause the 4508 to close the latch, and the chip will ignore any data at its inputs. In most cases the $\overline{store}$ input is kept low. When you want to store data in the latch, you get the data to the input pins and then send a short, positive-going pulse to the $\overline{store}$ input to open the latch and have the data stored in the chip.

Enable input The $\overline{enable}$ input on pin 3 is the control for the 4508's outputs, but it's important to realize that's all it does. If you bring this input high, the outputs will be tristated; making it low will allow the latch data to be present at the outputs. The state of this pin, and the subsequent state of the chip's output, has nothing whatsoever to do with the rest of the IC's operation. The other control inputs keep on doing their thing regardless of whether the 4508 outputs are tristated or not.

Other pins The other pins, power and ground, are at the standard locations on pins 24 and 12, respectively.

Now that we've decided what to use for storage, the next thing is to figure out what to do with it. This is really not too much of a problem. Because we want to store the data that's coming from the keycode encoder, the 4508 inputs have to be connected to the output of the 4518 that generates the data whenever we press a key on the keypad.

The design problem facing us at the moment is that we need a way to have the keyboard data appear at two places—the inputs of the 4508s so it can be stored, and the inputs of the 4511s so it can be displayed at the same time. If you study the circuit schematic we have so far, you can see that the display section—the 4511s and the LEDs—is a peripheral circuit. We've been talking about this section as if it were an integral part of the circuit, but it's really not.

The 4511s and the LEDs may be sitting on the same breadboard and may be an important feature when you use the circuit in some application or other, but they really don't have anything to do with making the whole keyboard circuit work. This section is a feature, not a necessity; and if you took it off the board, the circuit would happily continue generating keyboard code. You might not see it but, Zenlike, it would be there. Trust me.

The right way to think about the circuit we're building is to see it as a keyboard with the ability to store up to 10 digits and a display that displays the digits entered from the keyboard. But forget about the display for now and concentrate on the storage latches. Once we have a design that can store the entered data, we

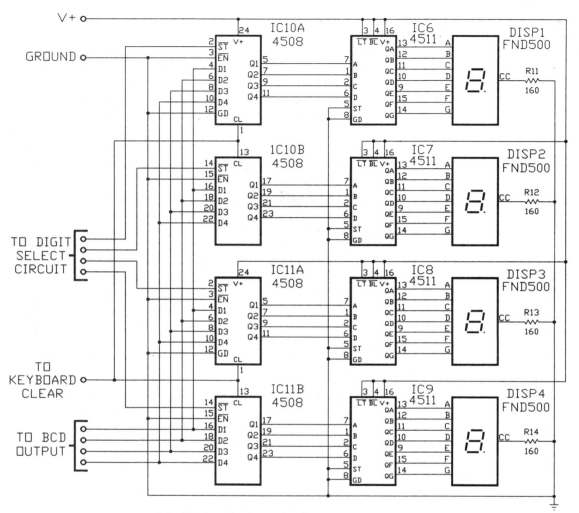

6-7 Keyboard schematic after adding the 4508s.

can start worrying about where to pick off the data for the display. You might have some ideas about that but put them aside for the moment.

It's a simple matter to connect the 4508 inputs to the output of the 4518 in exactly the same way we previously connected the 4511s that drove the LEDs. Refer to Fig. 6-7 and you'll notice that since the keyboard is generating 4-bit data, we're treating each 4508 as if it were two separate chips and using each half of the IC as an independently controlled latch.

As before, I'm only indicating the setup for four digits to make it easier to see, but if you want to expand it up to 10 digits, all you have to do is add the needed parts. The circuitry is exactly the same.

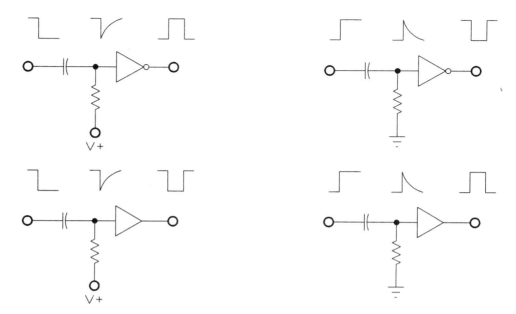

6-8 Half monostable pulse generators.

The digit selector we were using to control the 4511s can control the 4508s as well. Whenever we press a key on the keypad, we want the data to appear on the circuit's internal 4-bit bus. We also want the data to be stored in the appropriate latch. We still need the half monostables to generate control pulses for the latches; but if you think about it for a moment, you'll realize that we've just come up against a problem.

When we originally designed the half monostables, we configured them so they would generate a positive-going pulse (because the latch controls on the 4511s are active high). As you can see in Fig. 6-6, the latch controls for the 4508s are active low and that would appear to create a problem.

Not really. Half monostables, as shown in Fig. 6-8, are versatile circuits. As we discussed in the last chapter, they can be set to generate either positive- or negative-going pulses. From the illustration you can see that the only thing you have to do to change them is move the resistor from ground to Vcc. But before you consider this the end of it, think for a minute; and if you still don't get it, take another look at the drawing of the half monostables.

In order to get a negative-going pulse from the half monostables in the circuit, we have to do one of two things: either find some way to provide them with a negative-going input pulse or replace the inverters with straight buffers—something like the 4050.

It would be a shame to have to add more silicon to the board because a minimum of parts is a reasonable design habit to cultivate. There's no easy way to dump the 4049 because we need at least one on the board to invert the "any key pressed'' line and control the enable input of the 4518. This is to say nothing of the colossal pain in the neck it would probably be to rewire the breadboard. It certainly seems as if we should look at the alternative and see if there's some way to generate separate positive-going pulses whenever we press a key on the keypad.

Understanding the answer to this problem hinges on understanding how we made the circuit display work with negative-going pulses in the last chapter. If you don't have a good handle on that, take some time to go over it before we get into this.

The 4017-based digit selector works well, but one problem we had to overcome in the design was that there was no way to turn all the outputs off. The best we could do was generate a reset pulse at power-up so that we at least knew which output would be active. This still left us with a problem because the positive-going pulse generated at the output pin would enable the first 4511. The result was that the first time we pressed a key on the keypad, the second digit in the display would be enabled, not the first one.

The way around that was to connect the first digit in the display to the second output of the 4017. If you don't have that straight in your head, go over the discussion in the last chapter.

The way to generate a negative-going pulse for the store inputs of the 4508s without having to add or substitute parts in the circuit we already have (and suffer the attendant brain damage) is to realize the truth of the old saying

Whatever goes up must come down—even electronically.

Although it may take a while.

The first output of the 4017 in the digit selector goes high when the chip is reset at power-up, but it stays there until the first bit of data is entered from the keyboard. As soon as the key is pressed, the second output of the 4017 goes high and, more important, the first output goes low. Going low means that we have a negative-going pulse whenever a key is pressed, and that's exactly what we need to control the store inputs of the 4508s.

In order for us to modify the circuit we already have so it can be used to control the 4508s, we only have to make a few wiring changes on the breadboard. The resistors in the half monostables have to be connected to Vcc instead of ground, and the order of the 4017 outputs has to be changed. Instead of using the chip's first output as the control for the last digit, we can put things in a more logical order and have the first 4017 output control the first digit, the second output control the second, and so on.

It makes more sense this way because

Logic circuits should always be as logical as possible.

Whenever possible.

The way the circuit was configured, we were grabbing the state of the 4-bit keyboard bus as soon as the digit was selected by the 4017. The way things are now, we're latching the last thing on the bus before the next digit is selected.

Make sure you have a good understanding of what's happening here before going any further in the chapter. There's nothing tricky about the design, but you should carefully follow the kind of circuit analysis it took to come up with the answer. Being able to successfully design stuff at the bench means, as I've said 72 gazillion times before, being able to take a problem apart and examine each piece of it to find the right answer.

No matter what kind of goals you set for yourself, your success rate depends as much on how you think as on what you know. A good design is done on paper and in your mind long before it shows up as a working circuit on a breadboard.

I know some designers who don't like the occasional (?) brain damage that comes from doing the actual mechanical laying out of parts on the breadboard and only do the paperwork. These people are often referred to as systems analysts because, for one reason or another, they don't deal with the nitty-gritty of design details but prefer to see only the overall view of things. They work out the block diagram of the project and, once that's done, they farm out all the individual sections and concern themselves only with putting the pieces together and making the whole thing work. This is organization, not design.

Designing by committee may be OK if the end product is something like the Space Shuttle—so many different disciplines are involved that even a true Renaissance man like Leonardo would have a tough time doing everything. Most projects we do on the bench don't have quite the same level of complexity. Unless you've got something special on your mind, it's a lot more satisfying to work out the details by using your own mind instead of someone else's.

Final touches

Pressing a key on the keypad will now cause the data appearing on the bus to be stored in a 4508. As we continue to enter data, the digit selector will make sure that the next 4508 in line is ready to store the new data. Adding the latches to the circuit has taken care of one problem and, while we weren't looking, has given us the answer to another one of the design criteria we specified when we first began this project.

We now have a way to clear the bus from the keyboard—or at least we will as soon as we tie all the clear inputs of the 4508s together and route them to a common switch on the keypad. The only part of the clear circuit we haven't worked out is that it's also a good idea to reset the digit selector whenever the clear key is pressed. By doing that, not only is the data bus cleared, but the next key we press will cause the resulting data to be treated as the first keypress.

The final version of the keyboard circuit is shown in Fig. 6-9, and the placement diagram is in Fig. 6-10. The schematic includes several items that we haven't specifically talked about, but these are housekeeping details.

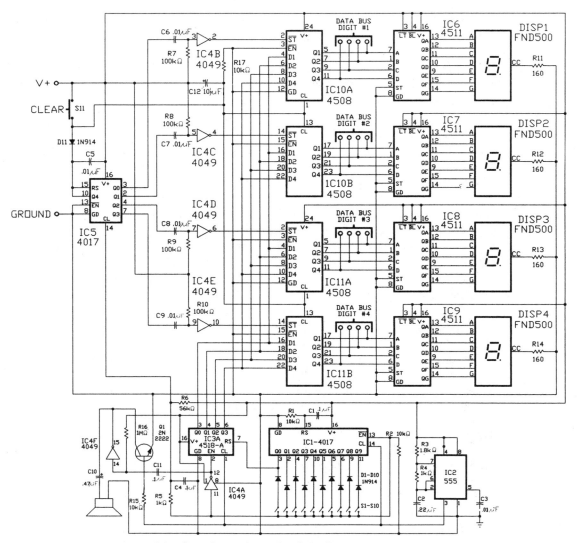

6-9 Complete schematic of the final keyboard circuit.

If you trace the keyboard clear design, you'll notice that the line is ordinarily held low by R17. This is standard stuff because we can't let the 4508 clear inputs float—simply a matter of good design. You should also note that a new diode, D11, has been added between the keyboard clear line and the reset pin of the digit selector. We want the clear signal from the keyboard to reset the digit selector, but we don't want the clear signal from the digit selector to reset the keyboard.

Don't forget that, as it's set up at the moment, the digit selector recycles through its digit count over and over again. After the last digit has been entered,

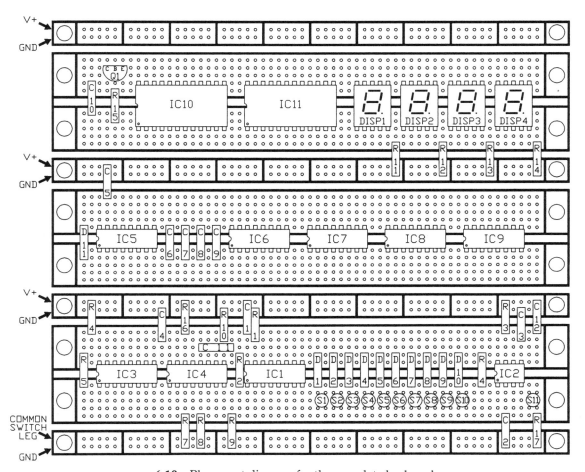

6-10 Placement diagram for the complete keyboard.

the 4017 in the digit selector resets to the first digit. This is a consequence of the design. When you use the circuit to its full limit—with 10 digits—a resistor should be added to tie the 4017's reset pin to ground after the power-up reset pulse is generated by C5. As shown in the schematic, a resistor isn't needed because the reset pin is being controlled by one of the outputs (pin 10). If you use all of the outputs to control digits, you have to add a resistor (exactly as we did with IC1) to keep the chip's reset line from floating (a CMOS no-no).

If you use fewer than 10 digits and have any one of the 4017 output pins controlling the chip's reset, remember that a reset pulse will be generated whenever the digit selector recycles. The diode, D11, makes sure that the reset pulse from the 4017 doesn't get through to the rest of the ICs controlled by the clear key.

You'll also notice that the output lines of the 4508s are now the output bus lines and that the display inputs are connected to this bus as well. As we said earlier, the display section of the circuit is a peripheral that monitors the state of the keyboard's external data bus exactly the same way as any other circuitry you might add. The store inputs of the 4511s have been tied permanently high because there's no longer any need to control their internal latches.

Whatever shows up on the external keyboard bus immediately shows up on the LEDs.

Oops!

If you're having problems with the circuit, finding the source of the trouble shouldn't be any harder than in earlier chapters. The way to go about troubleshooting is exactly the same as it was earlier—the fabulous wham-bam method.

We haven't really added that much in this chapter, and just about the only thing you might have that's causing trouble is a mistake in the lines controlling the 4508s. Do a quick check and be sure you remembered to tie the resistors on the half monostables to V+ instead of ground. And don't forget that we're not using the 4508's output enable inputs either so they have to be tied low. If you mess up and tie them high, the circuit might work but you'll be completely unaware of it because the outputs of the 4508s will tristate and what you'll see on the displays is anybody's guess.

If the circuit isn't beeping (and you're sure the beep generator is working correctly), you must have done something to one or more of the connections we made in the last chapter because we spent time there making sure the circuit beeped whenever a key was pressed.

Breadboards can get messy; and the more stuff you have wired up on one, the easier it is to accidentally pull one of the connections loose or cause the leads of two components, like resistors, to bend over and touch each other.

One easily made mistake is to miswire the clear key. It only takes two wire swaps to have a permanent high on the clear line and make the circuit hold eternal zeros on the output bus—an interesting setup but not a very useful one.

The keyboard circuit we've put together, as I noted in the last chapter, is a completely static one. As long as you keep your fingers off the keyboard, the only lines you should see moving are the outputs of the 555 keyboard clock and the 4017 keyswitch decoder. If you see movement on any other line, you've obviously identified one source of your problem.

I've seen lots of breadboarded circuits and, by far, the most usual source of trouble is mechanical rather than conceptual. This isn't to say that having only wiring errors makes things any better—a dead circuit is a dead circuit—but it does mean that you can find the wiring errors a lot easier. Just sit down with the paperwork and make sure that the only connections made on the board are the ones you're supposed to make.

Doing this kind of troubleshooting is only tedious for this circuit because you've got good paperwork and drawings to follow. There's a lesson here and it's well worth your while to learn it. All through these pages I've been talking about records, notes, and, in general, any kind of paperwork that can document the development of whatever it is you're building.

If it could be stressed more by writing it down in Neptunian, I'd write in that. If you don't produce a mound of paper every time you work on a project, the project isn't complete.

Maybe this will drive the point home: No matter how much talent you have as a designer, if you don't support your work with paper, you won't be able to keep a job—not even on Neptune.

Going further

Now that you have a working keyboard circuit in front of you, take a few seconds to cross off the rest of the boxes in the block diagram and then start figuring out what you can do to improve the way the circuit works.

Nothing is ever as good as it can be.

And nowhere is that more true than in the design we just finished.

Before you go nonlinear because I seem to be telling you that the keyboard circuit is incomplete, think about it for a minute. While the circuit does everything that we originally specified in the list of design criteria, that doesn't mean you can't redesign any of it to work differently, or even better (although that's more a matter of personal preference than anything else).

There are some improvements that I'd make to the circuit, but I'm leaving them to you. Some of them are interesting enough to be worth your time, and all of them are good exercises in design. It's up to you whether or not to do them; but if you've taken the circuit this far, it's a bit of a waste not to do some extra work on your own and take the circuit a bit further.

The first improvement I'd make is to change the way the digit selector works. As things stand now, it grabs the data from the bus for each digit just before the next digit is entered. This is fine in theory but, as the circuit is set up now, the feel of the keyboard is awkward because the data gets entered on the release of the key rather than on the press of the key.

There are several ways to correct this—some are easier, some are harder, and some are slicker than others.

The simplest way to handle the problem is to eliminate the reason the problem arose in the first place. I'm talking about replacing the 4508 latches. Remember that the reason the half monostables had to be modified was that they had to produce a negative-going pulse for the store inputs of the 4508s. If you spent some time with a series of databooks, you could undoubtedly come up with a latch that had an active high store input. This would let you put the digit selector back the way it was when we only had the display in the circuit and the digits were entered on the press of the key rather than on the release.

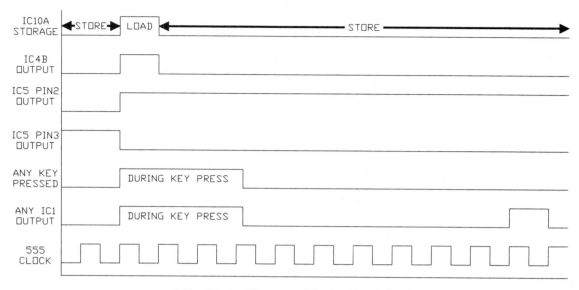

6-11 Timing diagram of the keyboard circuit.

You might figure out a way to build something that would pulse the digit selector when a key was pressed. As it is now, the 4017 (IC5) is clocked directly by the "any key pressed" line, and the width of the clock pulse is simply the length of time you keep the key pressed. If you use that line as the input for a circuit that produces a narrow clock pulse for the digit selector, the next digit will be selected so quickly that you won't notice anything wrong.

Remember that the data is being entered on the falling edge of the pulse from the digit selector's output, and that only happens when the digit selector moves to the next digit. If you don't have a clear grasp of this, look at the timing diagram in Fig. 6-11.

The real brute-force solution to the problem is to add a buffer like the 4050 to the circuit and use it to build the half monostables for the digit selector. We talked about this earlier and, if you're not bothered by adding some unnecessary silicon to the board, it's certainly a possible solution.

You'd still need an inverter to do the job currently being done by IC4A but, if you absolutely had to, you could build one out of a small transistor. And don't forget to redesign the half monostable that's currently controlling the beep generator.

A more basic design change would be to change the order in which the data is being shown. At the moment, the display is simply a way of looking at the data bus. It starts out with a full set of zeros; and as you load the latches with data, the number you entered shows up on the corresponding display.

It would be slicker—much slicker—to have the numbers appear on the display in calculator fashion as shown in Fig. 6-12.

The first digit entered would show up on the right-hand display; and when you entered the second number, it would overwrite the first one and the first number

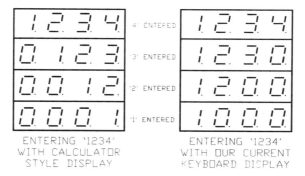

ENTERING '1234'
WITH CALCULATOR
STYLE DISPLAY

ENTERING '1234'
WITH OUR CURRENT
KEYBOARD DISPLAY

6-12 Current data entry order and calculator-style data entry.

would move over to the adjacent LED. I'm not going to tell you how to do this, but I will tell you that you can do it by simply rewiring the circuit you already have.

You'll need a way to separately trigger the latch and its related display, but the appropriate signals are already being generated by the circuit. You should be able to figure this one out for yourself if you review the discussion we had earlier about changing the triggering done by the 4017 digit detector, IC5.

It would be unbelievably cool if you figured out a way to take advantage of the blanking input of the 4511s. Nothing looks slicker than having the display stay blank until you start entering data. Leading zeros are a drag—they make the display look messy and confusing. All it takes is an extra bit of logic—from the circuit and from you.

When you actually use this keyboard circuit to do something like a real job in a real application, you might want to bring the enable lines of the 4508s out to the external data bus. Just tie them all together and make them available. The circuitry being fed by the keyboard might need to take the keyboard data bus off its input bus to prevent the dreaded condition of bus contention.

One interesting parallel between electronics and life as we know it is that you can't have two voices talking at the same time in the same place on the same line. In real life this leads to confusion, but in electronics it leads to smoke.

A last word

Any time you go through a design exercise like this one, you should take a few minutes when you're finished to understand what you got from it and what you get from it. The two of these aren't really the same thing at all.

What you got from building the keyboard is a circuit that will undoubtedly be useful in other designs you work on for either yourself or, hopefully, someone else. It's a good circuit to stick in your notebook so that when you draw a block diagram for something else you want to build, you can label one box with the words Keyboard Entry and know that you've already got that one safe in your back pocket (in a manner of speaking).

What you get from building the circuit is, hopefully, worth a lot more than the circuit you've built. You should walk away from the experience with an appreciation of how you go about designing anything in general, not something in particular. I'm talking here about the method to follow whenever you're starting a project. The procedure is to go from thought to paper to board. Trying to eliminate any of these steps means you'll be wasting lots of time on all the remaining steps—and you'll probably be dooming yourself to failure.

Now *dooming* is a strong word, and I'm using it for two important reasons. The first is that I want to really drive home the necessity to proceed logically through all the required steps for doing a design at the bench. There are lots of designers out there. Even though they work on different projects, the successful designers are those who do whatever they can to maximize their chances of success. That means paperwork and procedures.

The second reason is that I've always wanted to use the word *dooming* in a book.

Now that all this is behind us, it's time to turn our attention to something completely different (sound familiar?). Will it be good? Will it be interesting? Will it be great? We'll find out.

*A slick trick*_____

Most logic circuits have to be powered with a 5-V supply; thus, it's almost a standard practice to use some kind of voltage regulator in the design. I've drawn a circuit to warn against an overvoltage condition, but you can get serious damage from an undervoltage condition as well. This circuit is similar to the earlier one. By doing some rewiring, it will warn you whenever the system voltage falls below 5 V.

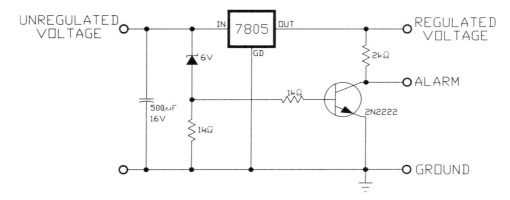

An undervoltage indicator.

Because the circuit is so small, you can add it to most of the circuits you build that use a three-terminal voltage regulator. By having it watch the input voltage that is going to the regulator, you can be warned of any problem with the unregulated supply. With the values shown, the transistor will conduct when the unregulated voltage has a value of less than 6.7 V.

7
Number crunching #1
Getting started

We've already seen that the amount of success you're going to have on the bench depends on a lot more than how much design experience you have under your belt. Knowing lots of stuff is great for things like games of "Trivial Pursuit" and TV game shows; but when you get right down to it, how many sets of matched luggage do you really need? Even Vanna White has some limits.

Having lots of electronic experience is definitely a plus whenever you're working on a project; but if you're loaded down with a bunch of bad habits, you're stacking the odds against success. Ask any project manager worth his salt what he wants in an employee and he's going to tell you that complete documentation and method are just as important as results—sometimes even more important.

When you're first faced with a set of design goals, your mind has to turn to paperwork, not boardwork. The steps we've been outlining in the previous chapters just can't be overlooked if you want to get anywhere at the bench. If a neat circuit idea pops into your head while you're still working on the design criteria, write it down and then put it out of your mind until you've reached the point in the design process where you're dealing with hardware.

Even if you've taken correspondence courses in memory training and can remember the middle name of your kindergarten teacher, you can't ignore the law that says

Nobody remembers everything,
and not everything is worth remembering.

The written word (including your notebook) will always be available whenever you want it.

You've got to be methodical about things or—and believe me, this is true—everything's going to get out of hand. You might actually be able to finish what you're working on by doing stuff in a nonorganized manner, but it's going to take you a long time to get to the end. And if you have to go back in six months or a year to make changes, just forget it. No notes means no looking back.

If you're like me, most of your work is solo stuff so there's no convenient human being around to bounce ideas off. This isn't always a bad thing, but you have to stay up on all the literature if you want to use the semiconductor industry's latest and greatest. Some of this stuff is useless because it's aimed at a particular market, some is priced beyond belief, and some of the new stuff is just unavailable. Remember,

If you can't afford to lose it, don't use it.

Breadboarding is really tough on silicon. One slip of the wrist and a working circuit can turn into a silicon junkyard. Not a pretty sight.

One consequence of the fact that new and terrific stuff is just about impossible to get in onesies is that people like us who work on our own generally like to stay with what we know and do our thing with parts we know we can get. If we need an oscillator in a circuit, we tend to look up oscillator in our notebooks and clone one of the schematics we have there. There's nothing wrong with this kind of design because an oscillator is an oscillator is an oscillator; and if it does the job—great. When we get to the hardware part of the process, most of the work is done with parts we already have around the bench. This is OK as well, but it does limit what kind of project you're willing to handle.

Don't read that last paragraph as any kind of official okeydokey to build circuits with junkbox stuff. You are what you eat and the same is true for board design. Junk breeds junk. Anyone who sits down to even prototype circuits with junkbox parts is opening the door to the possibility of serious brain damage.

There is nothing more difficult than designing and building a new circuit from scratch. At last count there are exactly 87 gazillion things that can go wrong, and your goal should be to do away with as many of those potential hassles as possible.

Breadboarding a new circuit with junkbox parts is a really stupid thing to do. We all know that the circuit won't work in the beginning, and finding out why can be a really brainbending experience. More than enough things can go wrong without also having to wonder if the parts you used were smoked before you put them on the board.

It's the nature of circuit development that stuff gets blown up on the board so who in their right mind wants to use stuff that might have been blown up before it was put on the board? You should have a sign on the wall above your bench that says

Avoid brain damage.

You're looking at lots of unavoidable hassles anyway.

The result of this unavoidable "I'll use what I know" attitude is that, besides not being up on the latest stuff (and producing circuits that may be compromises between what you want and what you know), you can tend to not use some really time-saving silicon. This is as true for exotic parts as it is for lesser-known standard parts. And by the term "standard part," I'm talking about your plain old, everyday, medium scale integration (MSI) chips.

A lot of the members of the CMOS and TTL families are designed for very specific jobs; and since the need for doing what these chips do doesn't come up very often, the chips aren't all that well known. In fact, most designers have come to think of MSI chips as being nothing more than a collection of counters, gates, buffers, and other basic building blocks whose only use is as a kind of electronic glue that holds more exotic parts together. Nothing could be further from the truth.

When looking through a list of either 4000-series CMOS or 74-series TTL chips, most people tend to only look at the familiar numbers—gates and such—but there are other, equally inexpensive and available ICs in these families that can easily solve some heavy circuit problems. You could build a circuit with familiar stuff that did the same job, but you'd be looking at acres of silicon and potential hassles.

It behooves you to get familiar with unfamiliar ICs because they can save a lot of brain damage—and besides, *behooves* is another word I've always wanted to use.

Doing arithmetic

If you're into designing electronics, I guarantee that sooner or later you're going to find yourself faced with a need for a circuit that can do electronic arithmetic. Everybody knows how to do division because one of the most common circuit activities is to divide master clocks down to needed frequencies, but that's not the kind of arithmetic I'm talking about here.

Doing frequency division is always a fixed-number application. You know the number you're dividing, and you've hardwired the number you want to divide by. There are several ways to do this, but all of them revolve around the use of a counter as shown in Fig. 7-1. A clock is fed to the input, and one or more of the outputs are decoded to pick off a particular number. Unless you're one of the chosen people or just out-and-out lucky, you usually have to gate some of the counter outputs together to get the number you want and generate the reset signal for the counter so the operation will repeat itself over and over again. This is pretty standard stuff. We do it all the time.

But let's suppose you have to do division where the circumstances are different. If the numbers you want to divide aren't constant and, worse yet, they vary unpredictably from moment to moment, there's no way you'll be able to use the standard counter approach to solving the problem. Oh, I suppose some of the

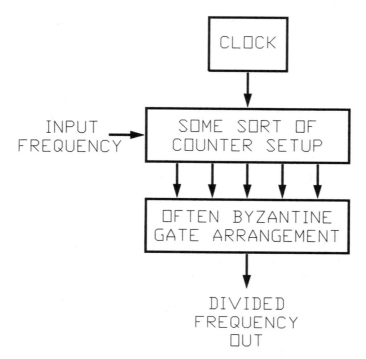

7-1 Standard method of frequency division.

adventurous out there could come up with a Rube Goldberg (a true genius since how many people can you think of whose name has become a common noun?) arrangement of gates and other stuff that would get the job done. I'd hate to troubleshoot it, though.

And just in case you're getting ready to do some mental gymnastics and design something like that, remember that I started this section by using the word

COMMON RATE MULTIPLIERS			
FAMILY	PART	TYPE	SIZE
TTL	7497	BINARY	64
TTL	74167	BCD	10
CMOS	4089	BINARY	16
CMOS	4527	BCD	10

7-2 Common rate multiplier chips.

arithmetic, and that word takes in a lot more than simple division. Arithmetic includes stuff like multiplication, square roots, and other operations that most of us haven't thought about for a while—with good reason. Who remembers how to manually get a cube root—or for that matter, even a square root?

Rate multipliers

It might surprise you to know that both the CMOS and TTL families have standard parts, called *rate multipliers*, designed specifically to do real arithmetic. The parts are listed in Fig. 7-2. You can see that they're available in either BCD or binary versions just exactly as we saw with some counters when we were starting the keyboard design. The only difference between them, as shown in the chart, is the maximum count they can have. But we'll get to that.

If you look these chips up in a databook, you'll see some slight differences between them but, in general, if you know how to use one of them, you know how to use all of them. After all, there's some rule somewhere that says ICs are designed to be easy to use, not difficult. Sure.

No matter what name the semiconductor manufacturer wants to give a rate multiplier, the most descriptive one is *number cruncher* because these chips can do a whole bunch of arithmetic with very little in the way of support circuitry. The 4089, typical of these ICs, is a CMOS binary rate multiplier. Even though you might never have run across it before, a quick trip to a parts catalog will show you that it's no more expensive than the other 4000 chips you know and love and is just as easy to find.

The pinouts for the 4089 are shown in Fig. 7-3. Although they might seem to be a bit peculiar, once we go through them you'll see that they're really very simi-

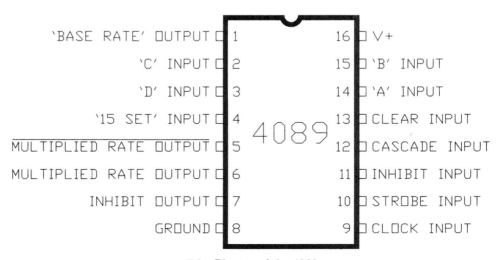

7-3 Pinouts of the 4089.

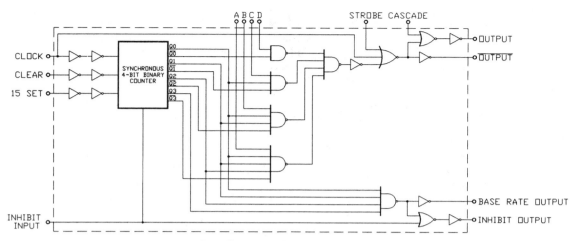

7-4 Block diagram of the 4089.

lar to a lot of the standard MSI stuff you use every day. There are inputs, outputs, and control pins just as with any other chip; and the only reason the names are unfamiliar is that the functions they have are unfamiliar.

Describing the functions of some of the pins on a rate multiplier is difficult to do clearly in words. As we go through each function, if something still seems screwy to you, take a look at the IC's block diagram in Fig. 7-4 and the chip's truth table in Fig. 7-5 to understand what's going on.

Don't be put off if you don't get a handle on everything about this chip right away. It's not that anything about it is so complicated as much as that a lot of it may be new. I'll bet that most of the pins on the chips you now treat as old friends were pretty weird when you first started using them as well. The rate multiplier may seem to be strange if you've never used one before (and most people haven't), but we'll put an end to that. Trust me.

Base rate output The base rate output on pin 1 carries a frequency equal to one-sixteenth (remember that this is a binary-based IC) the input clock at pin 9. The output here is one of the two frequencies we'll be using when we start doing arithmetic with the chip.

Data inputs The data inputs are, in a by-now-thoroughly-expected disorder, on pins 14, 15, 2, and 3. These are the binary-weighted A, B, C, and D inputs, respectively. When the chip is operating, the number put on these pins is one of the two numbers we want to multiply by.

"15 set" input The "15 set" input on pin 4 is one of several control inputs used to reset the chip and hold it with the outputs set to particular logic levels. The 4089 (like other rate multipliers) has more than one input for resetting so that the outputs can be frozen in more than one state. In normal operation this pin is held low.

Outputs The outputs are on pins 5 and 6. The frequency that shows up here is the second one (along with the base rate output from pin 1) that we'll use when we're doing arithmetic. The actual frequency you get at these pins is equal to the base rate multiplied by whatever binary number is presented to the four data inputs. At first glance, it would seem that the only difference between the two data outputs is that one is the complement of the other. The chip has an internal inverter that does this; and if your application only uses one rate multiplier, it doesn't matter which of the two outputs you use. The time when it does matter which output you choose is when you want to work with numbers larger than 16, the limit of 4-bit binary, and you cascade two or more chips together. This is because the pin 6 output is NOR'd together with one of the control pins. It's much easier to understand how the pins are arranged if you look at Fig. 7-4, the block diagram of the 4089, and study the IC's logic states shown in Fig. 7-5.

THE 4089 TRUTH TABLE

INPUTS AND PIN NUMBERS									OUTPUT PINS			
#14	#15	#2	#3	#11	#10	#12	#13	#4	#6	#5	#7	#1
A	B	C	D	IN	STR	CAS	CLR	SET	OUT	O̅U̅T̅	INH	BR
0	0	0	0						L	H		
1	0	0	0						1	1		
0	1	0	0						2	2		
1	1	0	0						3	3		
0	0	1	0						4	4		
1	0	1	0						5	5	ONE	
0	1	1	0		HELD LOW				6	6	PULSE	
1	1	1	0		FOR NORMAL				7	7	PER	
0	0	0	1		OPERATION				8	8	OUTPUT	
1	0	0	1						9	9	CYCLE	
0	1	0	1						10	10		
1	1	0	1						11	11		
0	0	1	1						12	12		
1	0	1	1						13	13		
0	1	1	1						14	14		
1	1	1	1						15	15		
DOESN'T MATTER				1	0	0	0	0	UNRELIABLE			
				0	1	0	0	0	L	H	1	1
				0	0	1	0	0	H	*	1	1
			1	0	0	0	1	0	16	16	H	L
			0	0	0	0	1	0	L	H	H	L
				0	0	0	0	1	L	H	L	H

* DEPENDS ON STATE OF BINARY INPUTS

7-5 Truth table of the 4089.

Inhibit output The inhibit output on pin 7 is used when two or more of the chips are cascaded together to allow arithmetic operations by more than 4 bits. Just as with the control inputs, the 4089 (as well as all the other rate multipliers) has several control output signals such as this one. The reason for this diversity, as we'll see later, is that there are several ways in which rate multipliers can be cascaded and each has its advantages and disadvantages.

Clock input The clock input is on pin 9, and it wants to see the same standard square-wave frequencies fed to any CMOS clock input. Contrary to what you might have assumed, the frequency fed to this pin has nothing to do with the arithmetic operations of the IC. The only effect the clock rate has on the chip is to determine how fast the chip is going to do arithmetic for us. This may seem confusing; but when we start to do some circuit design, the confusion will disappear.

Strobe input The strobe input on pin 10 is a control for the outputs on pins 5 and 6. If you make this pin high, the rate multiplier will still operate internally but the outputs will be frozen at the particular state shown in Fig. 7-5. In normal operation this input is held low; but during certain circumstances when two or more chips are being cascaded together, this input is clocked by external signals. If you have no idea what this means, don't worry because we'll get into it as soon as we start designing circuitry.

Inhibit input The inhibit input on pin 11 is used along with the strobe input to control the 4089 when two or more chips are cascaded together. During normal operation this input is kept low. Don't be misled by the name of this pin because it's not really an inhibit input in the sense that it will safely halt the operations of the chip. If you make this pin high, the chip will stop working but the state of the outputs will be completely indeterminate and depend on the state of the chip's internal counter at the exact moment you made the inhibit input high.

Cascade input The cascade input on pin 12 is used, as you probably guessed, when two or more rate multipliers are cascaded together. A glance at the 4089 block diagram (Fig. 7-4) will show you that this is a gate control for one of the outputs. The truth table (Fig. 7-5) shows you that putting a high here will, as with all the other control inputs, cause the chip outputs to freeze; however, two of them (the base rate and the inhibit outputs) will continue to operate normally. Like the rest of the control pins on the 4089, the cascade input isn't designed to be used by itself. It's intended to be used together with other inputs and outputs to control the chip's operation whenever several rate multipliers are cascaded together.

Clear input The clear input on pin 13 is, finally, a control pin whose name is an accurate reflection of its function—sort of. Bringing this pin high will reset the chip's internal counter and freeze the output pins at a predetermined logic state. This is shown in the truth table (Fig. 7-5), but you'll also see that the "sort of" comes from the fact that the state of the pin 5 and pin 6 data outputs will depend on the state of one of the data input pins—more precisely, the state of the D input on pin 3. This is important to keep in mind because it means that when you want to use the clear input, you should gate it with the D input so you know exactly what to expect at the outputs.

Other pins The other pins are standard power (pin 16) and ground (pin 8); and, believe it or not, there's nothing kinky about these pins at all.

There's no doubt in my mind that this stuff has done a staggering amount of brainbashing. Anyone who tries to understand how a chip operates by reading a description of the pins is guaranteed to suffer an extraordinary amount of brain damage and, worse yet, get absolutely nothing of value for the experience. It's important to have that kind of material around for reference, but trying to make sense out of it is only slightly more ridiculous than trying to remember it.

This kind of information is important to have when you have a bit of familiarity with the IC but, by itself, it won't get you familiar with the chip at all. The more time you spend at the bench hunched over a breadboard, the more you realize that there's no similarity whatsoever between theoretical and practical knowledge—and nowhere is this more true than with the 4089 or any of the other rate multipliers.

The simple explanation of how this chip operates is that you feed a clock to pin 9, ground most of the control pins, put a binary number on the data inputs, and the chip will give you two different outputs, both of which are the principal reason for using the chip in the first place.

The first output on pin 1 is the *base rate*, which is simply the input clock divided by 16. Another way of saying this, and one that helps a bit more to understand the terrific stuff this chip can do for you, is that you get one pulse at the base rate output for each 16 pulses at the clock input.

The second output is at pin 6 (with an inverted version appearing at pin 5). And for want of a better name, we'll call it the *multiplied output*. This is the output that makes learning about the rate multiplier worth whatever amount of brain damage it takes. The signal that comes out at this pin is equal to the base rate multiplied by the number at the binary inputs. Another way of saying this is that for each pulse at the base rate output, you get a number of pulses that's the same as the number at the binary inputs.

If this isn't clear, let's take an example and say that we have a 6 present at the binary inputs, we're feeding a 16-kHz frequency to the clock input at pin 9, and all the control pins are tied wherever they're supposed to be—a basic rate multiplier setup.

Once we apply power to the chip and look at the outputs, we'll find we've got a 1-kHz frequency at the base rate output (one-sixteenth the input clock) and a frequency of 6 kHz at the multiplied output (the base rate times the binary number). Just the way the databook (and you do have one on order, right?) says it's supposed to be.

Even though you may not yet have a handle on how this chip is going to do arithmetic, several other possibilities should be sneaking into your mind. I mean, take a second to fully understand what we just did. We converted a number into a frequency. All we had to do to get a 6-kHz clock was put a binary 6 on the chip's data inputs and pick the 6-kHz frequency off the multiplied rate output. Not only that, but if we wanted a 12-kHz frequency, all we would have to do is either change the number at the inputs or double the input clock. If the only change we made to this example was to up the input clock to 32 kHz, we'd get a 2-kHz frequency at the base rate output and 12 kHz at the multiplied rate output.

Is this great or what? Sorry, people, but it's "or what."

All this stuff seems terrific, and sometimes it is, but it's also as completely misleading as a TV ad for aspirins. Sure, three out of four doctors recommend Head Helper aspirins, but it's also true that the Head Helper Company had to call 80 gazillion doctors to find three who used their product. While it's not exactly lying to claim that three out of four doctors recommend Head Helpers, it's a lot more accurate to say that three out of 80 gazillion doctors recommend Head Helpers. Of course that doesn't sell anywhere near as many aspirins.

To see how things can get messy with the 4089 (or any other rate multiplier), let's suppose we replace our very carefully chosen 16-kHz input clock with something else—say, 23 kHz.

The base rate is one-sixteenth the clock input; thus, with a 23-kHz clock, the base rate frequency would be 23 kHz divided by 16, or the somewhat mysterious rate of 1 437.5 Hz. Before you try and figure out what half a hertz means, imagine that we had a 9 at the binary inputs. The multiplied rate output would be nine times the base rate or 12 937.5 Hz—an equally strange number.

But let's suppose we do have numbers like the ones we used in our first example—nice even ones that produce nice even results. Even though nothing like this ever happens when you're dealing in the real world, we can still look at it as some sort of ideal situation that's going to produce ideal results. Don't count on it.

The 4089 is a complex chip internally, and one of its components is a synchronous counter used to do the divide-by-16 calculation for producing the base rate. The counter's outputs are also used to feed the logic needed to make sure the correct number of pulses show up at both of the chip's multiplied rate outputs.

One consequence of using a synchronous counter (as opposed to a ripple counter) is that clock pulses at the chip outputs always are in sync with the clock on pin 9. Depending on whether you look at the pin 5 output or the complementary output at pin 6, the rising or falling edges of the output pulses are going to coincide with the rising edge of the input clock.

The next thing to realize is that, since the rate multiplier's outputs are gated on and off by internal logic that's clocked by the chip's input clock, these output pulses are always going to be exactly one input clock pulse wide. This is just the way the chip works.

Keeping both these facts in mind, let's go back to the ideal numbers we used when we started this discussion. Just in case you've forgotten, that means an input clock of 16 kHz and a 6 at the binary inputs.

When we set the chip up like this, we're going to get six output pulses for each 16 input clock pulses. We already saw how this happens; but with our new understanding of the chip's innards, let's look a bit more closely at exactly what the chip is going to give us at the outputs.

The numbers in our example work out nice and evenly (none of those weird half-a-hertz numbers), but the output on the 4089 is anything but the kind of wave-

PIN	A	B	C	D
STATE	L	H	H	L

PIN #6
OUTPUT

PIN #9
CLOCK

7-6   Theoretical timing diagram of the 4089.

form you'd want to use for timing or feed to most other clock inputs. We're getting six output pulses for each 16 input pulses; and while the arithmetic works out evenly, we also have to remember what the makeup of the 4089's internal circuitry does to the output pulses. They're in sync with the clock pulses and have the same width.

What this does to the output clock from the 4089 is make it look a lot like a broken comb. While all the teeth are the same size, the spacing is uneven. And this is really the way it is. Not only that, but if you think about it for a second, you'll realize that there's no way you can get nice 50% duty cycle type square waves out of the 4089 unless you want 16 output pulses for each 16 input pulses.

This may be a bit hard to visualize so take a look at the ideal output waveforms in Fig. 7-6. Now I'm the first one to admit that there's a big difference between ideal drawings and what you'll see if you put a scope on a working circuit, but in this particular case the ideal drawings are right on the money. All you have to do is compare the ideal waveforms to the ones shown in Fig. 7-7, the real waveforms I got by setting the 4089 up to do the same multiply-by-6 operation we've been using as an example.

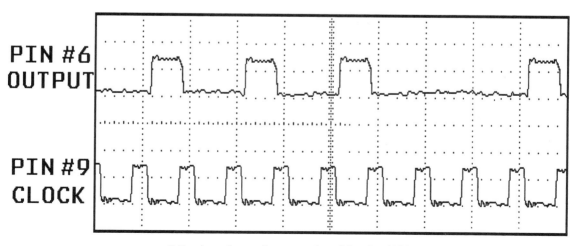

PIN #6
OUTPUT

PIN #9
CLOCK

7-7   Actual waveforms produced by the 4089.

Now that we've gone through all this, the time has come to face the painful truth of using rate multipliers. It's unfortunately true and inescapable that

You never get exactly what you want.

Uneven pulses and output shakiness are built-in consequences of the way rate multipliers work. But don't start demanding your money back just yet.

Even though the output waveforms of the 4089 and of all the other rate multipliers won't win any awards for beauty or symmetry, the way they work and the job they do make them ideal for any application where all you want is a total number, not a constant frequency. Working toward a total number is, as I seem to remember, one of the goals of doing arithmetic.

Before we start doing some design work, there's one more thing to say about the idea of using a rate multiplier for doing down-and-dirty frequency division. Even though the output waveforms are going to be terrible (and I wouldn't want to use them as a metronome for practicing the piano), the circuit is so versatile and easy to do that I have to admit the idea of ignoring the disadvantages and doing it anyway is really very tempting.

There's absolutely nothing stopping you from getting a handful of rate multipliers, connecting them to a clock and a rotary switch, and doing some dial-up frequency division. The outputs will undoubtedly be really messy, but you can't deny that they'll be correct—and that's something. Not much, but something.

As I've said many times before, this is America and you're free to do what you want with what you buy. If you decide you want to use a rate multiplier for frequency division, that's your choice, but don't do any critical work with the answers you get from the circuit. This is the kind of shortsightedness that leads to making bridges that fall down and tunnels that don't meet in the middle. It's a quick and easy shortcut that bypasses a lot of work but

Shortcuts are faster but you miss seeing stuff.

Sometimes there are things you can't afford to overlook.

Here's a great story and it's even true.

A few years ago, I found a neat-looking brass-and-wood antique tool at a flea market. It was like an outside measuring micrometer, and the markings on the wood body and brass slide indicated that it was used for measuring the diameter of rope. You would clamp it on a piece of rope and measure the diameter—inches were etched on one side of the slide and circumferences were etched on the other. Once you had the number, you could get the weight of different lengths of various kinds of rope from a table inscribed on the body of the tool.

It took me a while to realize that there was something wrong with the way the numbers for calculating the diameter were etched on the slide. From how the numbers lined up, circumferences on one side of the slide and inches on the other, it was clear that the people who made the tool were using 3.00 as a value for pi (π).

I checked this with a friend of mine who, among other things, is an authority on rope (he works as a consultant for the Navy); and he told me that at the turn of the century, assigning a value of 3.00 to π was the accepted practice.

It's not too bad in the short run but . . ..

Someone else told me that the Indiana state legislature once passed a law declaring that π was equal to 3 because the speaker of the house had a hard time doing division. I can't swear that this really happened, but it's certainly the kind of thing I'd expect a politician to do—at least until bridges and stuff started falling down.

Defining design criteria

Whenever you set out to design something, you have to follow steps, or standard procedures, that take you from the first inkling of an idea to mass production in Taiwan. You should know all this stuff by now—you better—because I've been telling it to you over and over again.

These rules, like any rules, occasionally have to be modified when circumstances dictate. Our goal now is to design a circuit that will allow us to do arithmetic with the rate multiplier—and that's cool. But before we do that, it's a good idea to get a bit of practical experience with the chip so we can see how it's supposed to work. Knowing how things should be when they're working correctly is a good basis for finding the problem when things stop working or don't work at all.

Rate multipliers aren't all that difficult to use; but if you've never played around with them, setting them up to do their thing can be somewhat confusing. Since I'm a firm believer in learning by doing, the first thing we're going to do with the 4089 is design a circuit that demonstrates the chip's basic operation.

Even though we're just designing a demonstration circuit, we still have to follow the general design guidelines for project development that we've been using for everything we've done so far. In a nutshell there are three steps to a completed design:

1. Brainwork
2. Paperwork
3. Boardwork

And as I've said 63 gazillion times before this, they have to be done in that order.

The simplest thing we can ask a rate multiplier to do is some plain two-number multiplication. Doing that will show you just about all the basic principles of designing with a rate multiplier.

Now that we have an idea of what we want to do, the first step in the design (and the first step in any design) is to draw up a list of design criteria. Since the circuit we're building is relatively limited, our list of criteria isn't going to be very long.

You can add to it if you want, but keep in mind that what we're doing here is only going to be a step toward a more complex and useful circuit for doing arithmetic. It's enough work to successfully design what you need without the extra burden of having to design something you really don't need. There are, after all, limits.

1. The circuit will multiply two 4-bit numbers together.

We can do 4-bit multiplication with just one rate multiplier. While real-world applications will undoubtedly require the ability to handle larger numbers, it's always a good idea to start off with just the basics and then build on that. Even this number-size restriction isn't that limiting. As we'll see when we get into the details, only one of the numbers has to be less than 16 (the limit on a 4-bit number).

2. The result of the multiplication will be displayed on LEDs.

This really isn't necessary to make the circuit work, but it's a nice convenience. It's also a bit of planning for the future because a full-blown arithmetic circuit (as opposed to this somewhat limited one) will need a display anyway.

3. All the parts will be both cheap and available.

Unless you're working for the government or have really deep pockets, this should always be on your list of design criteria because there's always a high silicon casualty rate when you're doing a prototype on a breadboard. Some parts are hard to find, and nobody on Earth wants to be forced to get parts from Klant Zorch Electronics—even if it's the biggest company on Neptune.

We could go on with the list, but that's not really necessary just now because this is enough for the circuit we're designing. In any event, we'll be adding to the list when we expand the circuit into something a bit more useful.

Block diagram

We've already seen that doing a block diagram is where you get an idea of the circuit elements you'll need to complete your project. In this case it does even more because it will show you exactly how the rate multiplier works. Remember that the more complex the IC, the more it needs in the way of support circuitry to work.

The block diagram for the rate multiplier demonstration circuit is shown in Fig. 7-8. Although some of the sections may seem to be new, we've actually talked about all of them already. Before we start to talk about the hardware, let's do a quick review of the basics of rate multiplier operation.

A rate multiplier takes a clock and a 4-bit number and uses them to give you two different outputs. The first output, in the case of a binary chip like the 4089, is the base rate and that's equal to one-sixteenth of the clock frequency. The second output is the multiplied rate and that's equal to the base rate multiplied by the binary number present at the chip's inputs. What makes it possible to do arithmetic with this chip is the relationship between the two outputs.

Simply stated, for every pulse at the base rate output, the number of multiplied rate pulses you get is the same as the number present at the chip's data inputs.

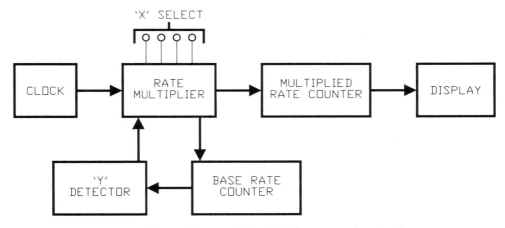

7-8 Block diagram of the 4089 demonstration circuit.

In order to multiply two numbers, X and Y, with a rate multiplier, we have to put one of them at the chip's data inputs and then start the clock going. As soon as we do this, we have to keep track of both the base rate and the multiplied rate to get our answer. The rate multiplier gives us the answer by counting up to X over and over at the multiplied output rate and telling us how often it's doing it at the base rate output.

If we wanted to multiply 5×6, for example, we would put 5 at the chip's data inputs and start the clock. For each five pulses we get at the multiplied rate output, we get one pulse at the base rate output. To get the answer to 5×6, we would build a circuit that counts the multiplied rate pulses, keeps track of the base rate, and stops the chip after six base rate pulses have been detected. This is exactly the kind of circuit shown in the block diagram.

There are two counters in the block diagram just as we described in the last paragraph. The base rate counter keeps track of the number of times the multiplied rate gets sent to the multiplied rate counter (and on to the display). As soon as the Y detector indicates that the rate multiplier has counted to X a predetermined number of times (by counting the pulses from the base rate output), it sends a stop signal to the rate multiplier.

Now that we've gone through the operation of the rate multiplier, you should have a handle on just how the chip does multiplication. In a real sense, the rate multiplier is really doing successive addition rather than some electronic equivalent of the times tables. When you ask it to multiply two numbers together, it repeatedly counts up to one of them and gives you a way to track the number of times it does it.

The clock shown in the block diagram can be anything you want, from a 555-based circuit to a simple oscillator built out of a handful of convenient gates. Because we're using a 4089 CMOS rate multiplier, the clock requirements are pretty loose. If you use one of the TTL chips in the rate multiplier family, you'll have to make sure that the clock you choose generates pulses that are TTL compatible.

The next thing to consider for the clock is just what frequency we should use. Whenever you have to design a clock for use in a circuit, there's always some Goldilocks frequency you should be aiming for. It can be because of the requirements of the parts you want to drive with the clock, limits set by what the circuit has to do, or even restrictions dictated by how the circuit is going to be used. A rate multiplier clock is somewhat unusual. The frequency doesn't matter—an amazing statement, huh?

Once you understand this, you'll know that you're well on your way toward having a real understanding of how rate multipliers work. The best way to see this is to deal with it mathematically. If we wrote down all the relationships between the frequencies associated with a rate multiplier as a series of formulas (or, as they like to say in the technical journals, formulae), we'd have two basic equations:

$$\text{Base Rate} = \text{Input Clock/16}$$

$$\text{Multiplied Rate} = (\text{Input Clock/16}) (X)$$

where X is the number we've put in binary form at the data inputs of the 4089. Each time we get one pulse at the base rate output, we get X number of pulses at the multiplied rate output. When we multiply X by another number, Y, all we're doing is letting the rate multiplier count up to $X\ Y$ times. It's just successive addition. Take a look at the same thing in mathematical terms:

$$\text{Multiplied Rate} = (X) (\text{Base Rate})$$

so
$$X = \frac{\text{Multiplied Rate}}{\text{Base Rate}}$$

but
$$\text{Base Rate} = \text{Input Clock/16}$$

and
$$\text{Multiplied Rate} = (\text{Input Clock/16}) (X)$$

so

$$X = \frac{(\text{Input Clock/16}) (X)}{\text{Input Clock/16}}$$

or
$$X = X$$

What this bit of symbolic arithmetic shows you is that of all the numbers fed to the rate multiplier, the only one that has any direct effect on the number you'll get at the multiplied rate output is the number you put on the data inputs—and they're the same! More than that, they're the same regardless of the frequency of the input clock or the kind of internal division done by the rate multiplier. The 4089 works on a base of 16, but that's none of our business. As far as using the chip in a circuit goes, for all we care it can do its internal work as binary numbers, decimal numbers, or even imaginary numbers (used by the fabulous fish-faced people of Neptune).

When you're working with rate multipliers, the only numbers you have to worry about are the numbers you want to multiply. Remember that.

When you're considering the choice of clock frequency for the rate multiplier, you should really use the highest one your circuit can tolerate. Because the circuit we're talking about building now is only going to be using one rate multiplier, we're somewhat limited in the size of the numbers we can multiply. Real-world circuitry using rate multipliers will have as many chips as are needed to handle whatever size numbers it has to handle. The calculations can get rather large so the best decision is always to have a fast clock when you're using a rate multiplier.

Schematic

Putting a rate multiplier on a breadboard and making it do actual work is fairly straightforward. The chip doesn't need an overwhelming number of support circuits, thus its basic operation can be demonstrated with a minimum of boardwork and a fair amount of brainwork. All you need is a clock, some counters, and a little bit of logical silicon glue to hold everything together. Lay the parts out on the breadboard as shown in Fig. 7-9.

Most rate multiplier applications use the fastest clock they can to get the answers as quickly as possible. Our demonstration circuit is a bit different. We won't be doing anything with it that has to be done in a hurry so we can pick any value we want. The clock I'm using in the circuit is based around a 555 (refer to the schematic in Fig. 7-10), and the clock frequency is up there around 50 kHz. You can use this or, if you have a problem with it for some reason, you can use anything else that produces well-shaped, CMOS-compatible square waves.

Now that we know more than you probably thought there was to know about the 4089's clock requirements, we can turn our attention to the only other part of the circuit you might have questions about—the base rate counter and the Y detec-

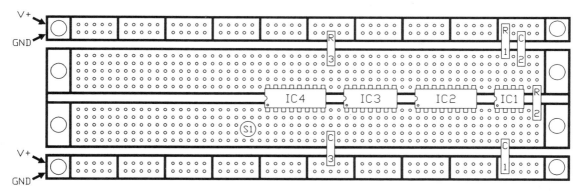

7-9 Placement diagram for the 4089 demonstration circuit.

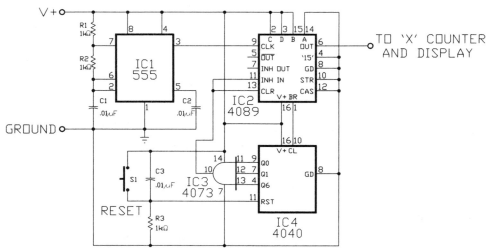

7-10   Schematic of the 4089 demonstration circuit.

tor. Before you get upset about it, take a look at the schematic and you'll see that all it takes to build this section is a counter and some gating.

I'm using a 4040 as the base rate counter because it's the perfect single-chip answer to having something that can count up as far as 4096. Because the base rate counter is also where you use one of the two numbers you're multiplying together, it's a good idea to have as long a counter as possible. Numbers larger than 4096 will require another 4040 (or at least a different counter arrangement), but this is more than adequate for now.

You've undoubtedly noticed that I've hardwired the multiplication problem into the schematic. The 4089 inputs are set to see a binary 14 (A = 0, B = 1, C = 1, and D = 1); and the 4073 output of the three-input AND gate will go high when the 4040 gets up to a count of 67.

The 4040 has consecutive binary outputs (Fig. 7-11) that go from Q0 to Q11; and I'm using the AND gate to decode Q0, Q1, and Q6. As soon as a count is reached in the 4040 where those three outputs are high, a high will appear at the output of the AND gate, be sent to the 4089's inhibit input, and then go on to reset the 4040 as well. This action will both disable the rate multiplier and clear it at the same time.

Using the circuit is simple. When you turn it on, the 4040 will be reset to a zero count by the brief positive pulse generated by R3 and C3. Once that happens, the outputs of the 4040 will be low, causing the output of the 4073 to go low. This will enable the rate multiplier, which will start responding to the clock pulses from the 555. Since a 14 is being presented to its data inputs, the rate multiplier will start sending 14 pulses out its multiplied rate output for each single pulse sent out its base rate output.

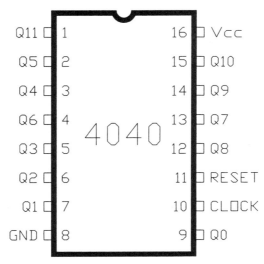

7-11 Pinouts of the 4040.

The 4040 will start counting the base rate output pulses and, when 67 of them have been received, the Q0, Q1, and Q6 outputs will be high. This will be decoded by the 4073, its output will go high, and that will disable and clear the 4089. The circuit will stay in exactly that condition until you press S1 to reset the 4040, clear its outputs, and cause the cycle to begin all over again.

You should notice that when the 4073 detects a 67 count at the 4040 and disables the 4089, the whole circuit doesn't stop dead in its tracks—the clock is still putting out pulses. We're only using this circuit to demonstrate the basic setup and operation of the rate multiplier; thus, it's not too important to disable the entire circuit at the end of a counting cycle. In real-world circuits it's often a different story.

If you want to spend the brain time, it's a good exercise in design to control the 555 along with the 4089. There's no mystery about how to do this—we discussed it a couple of chapters back when we were just getting started with the keyboard circuit. Stopping the 555 is simply a matter of taking control of its reset input at pin 4.

The schematic in Fig. 7-10 shows that this pin is tied high (as it should be when you want the 555 to run); and if we want to disable it, we have to bring it low. When we were building the keyboard circuit, we had a handy signal to use that went low at exactly the right moment to disable the 555. This time around, we don't have anything like that because the output of the 4073 is low when the circuit is working and high when it's not—exactly backwards.

We could take care of this by putting an inverter on the board, but instead of using an entire IC, it's easier to build an inverter with a small signal npn transistor as shown in Fig. 7-12. When the output of the 4073 is low, the transistor will be in cutoff and a high will be present on pin 4 of the 555, allowing it to run. When the

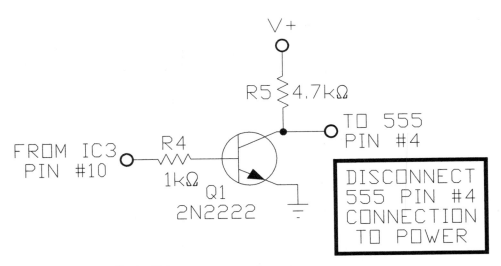

7-12   Using a transistor to stop the 555-based clock.

output of the 4073 goes high, the transistor will be turned on and the 555's reset pin will be low, disabling the clock.

Pressing S1 will enable the clock along with the 4089 and the rest of the circuit.

The last piece of business to take care of in the circuit is the display, and I haven't said anything about it because it doesn't really have anything to do with rate multipliers. We did include it in the list of design criteria, but it's more of a feature than a necessity.

It's a bad idea to add the display to the circuit unless you have a thorough understanding of all the rate multiplier fundamentals we've been talking about here. While the chip's basic operation is not very complicated, things start to get a bit more complex when you begin cascading the rate multipliers; and there's just no way you're going to be able to understand any of that until you've got all the material in this chapter down cold.

Oops!

If you're having any trouble with the circuit, it has to be something simple and mechanical because there aren't many parts or connections. Check everything against the schematic of Fig. 7-10 and be on the lookout for shorts. They can sneak up on you.

One handy way to check the circuit's operation is to slow the clock down so you can actually see the pulses on an LED. Put an LED at both the multiplied rate and base rate output pins of the 4089 and connect them to ground through 1-kΩ resistors. Replace C1 with a 1-μF capacitor to slow the clock down to something that lets you see each individual pulse at the outputs of the 4089.

If you have a scope, you can watch the 4089 outputs while they're running at full speed. Putting the clock at a crawl and watching the LEDs, however, not only lets you check things out but also lets you see just how uneven the multiplied rate outputs are.

When the circuit is operating correctly, you'll be ready to add the display, cascade the rate multipliers, and do a bunch of other really neat stuff with them, so get to work.

I'll be waiting for you.

*A slick trick*_____

Every once in a while you need some way to generate sine waves with a minimum of parts. This circuit will output low-distortion sine waves as long as you keep the output going to high-impedance inputs (better than 50 kΩ) and use components in the 741 feedback loop with measured values that are within a very small percentage of each other. The frequency of the sine wave can be calculated by plugging the component values in the formula

$$F = \frac{1}{2}(\pi)(RC).$$

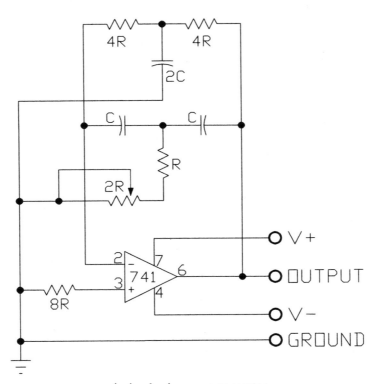

A simple sine-wave generator.

The circuit is self-starting; and if you're careful about the components you use and what you do with the output, you'll find the circuit to be both reliable and cheap to build with easily available parts.

8
Number crunching #2
Bigger buses & meatier math

Nothing is ever perfect. People have been arguing for centuries about what the word *perfect* means—wars have been fought about it—but no matter who does the yelling and screaming, nobody's ever been able to figure out exactly what the word *perfect* really means. Well, after seconds of scholarly research, I'm here to tell you that I not only know what the answer is, but I'm going to share it with all of you.

The simple truth is that

Perfection means good enough.

Anyone who doesn't agree with that can step outside and meet me around the back.

Electronic design is a personal business. Given certain goals, there will be as many perfect answers to a problem as there are people working on them. Nobody sees the same thing the same way; and the more complex the goal, the more ways there are to achieve it. The possible variations multiply geometrically.

Even simple circuits like the one we built to demonstrate how the rate multiplier works can be done a number of ways. The minute you find that gates have to be added to a circuit, you open the door to solutions based on logical thought; and that means the answer you get will depend on how you think. If you're a positive person, you'll use AND, OR and other positive logic. If you're not, all your designs will tend to be populated with NAND and NOR gates. One isn't better than the other—it's just different.

Turning the demonstration circuit we just did into something that can be used in the real world means adding a degree of complexity to it that's not difficult to understand but does leave room for lots of possible solutions. Even the chip itself, as we'll see in a minute, can be cascaded in two different ways. Which one you

choose depends, of course, on what you want to do; but it also will be a reflection of
how you think about things.

Unfinished business—the display

Before we get down to business and list the additional criteria for the additions
we're going to make to the circuit, we have to finish designing the display we called
for in the last chapter. Since we specified that we were going to use LEDs—I like
them a lot better than liquid crystal displays (LCDs)—we need a display driver. The
4511 we used for the keyboard is as perfect a choice here as it was there. To keep
you from having to flip back through the pages, the pinouts of the 4511 are shown
in Fig. 8-1. Your databook is on order, right?

The only difference between the display requirements for the rate multiplier
circuit and the ones we had for the keyboard is that we now need a display that can
handle the number of digits we're expecting in the answer we'll be getting from
the multiplication. In this case we're multiplying 14 × 67; and since this is the
planet Earth, we better get an answer of 938—and that means we need a 3-digit
display.

You could use the same display setup we had for the keyboard, but you'll run
into trouble trying to cascade 4511s because they're not really designed for it.
There's just no getting around the fact that a three-digit display is going to need
more than three drivers and three LED displays. You're going to need more silicon.

The 4553 shown in Fig. 8-2 is a one-chip answer to the problem because it has
an internal counter that can go up to 999. This is a useful IC to get familiar with.

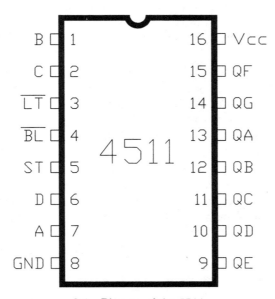

8-1   Pinouts of the 4511.

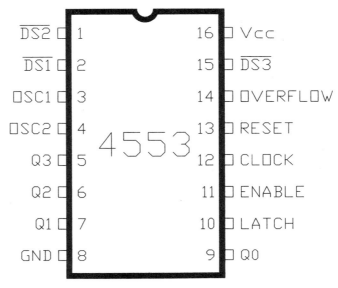

8-2 Pinouts of the 4553.

Once upon a time when Radio Shack was big on electronics (rather than consumer electronics), you could find this chip blister-packed on the shelf. As things stand now, it's still readily available from mail-order houses, but you can't help occasionally looking back to the good old days and wondering why it's all changed.

Be that as it may, the 4553 is a terrific chip. It generates the control signals for cascading, handles the display multiplexing of up to three digits, and even has its own internal latch. Let's take one of our by-now-famous trips through the pins.

Clock input   The clock input is on pin 12. As with all the other chips we've been using, this CMOS input wants to see a clean square wave with sharp shoulders—going up and coming down. Although this isn't really a consideration at the moment, you should make a permanent note to yourself that the 4553 advances its counter on the negative-going edge of the input clock. We'll be feeding the chip with complete pulses so there's no problem using it with the rate multipliers. However, if you want to use the 4553 somewhere else, this could be an important thing to be aware of.

Latch enable   The latch enable control is on pin 10. Since this chip has its own internal latch, it follows naturally that there's a control pin to open and close the latch. When this pin is low, all the clock pulses coming into the chip are added to the total stored in the latch, and the resulting number shows up at the chip's outputs. Making this pin high will close the latch doors. While the count will continue to add up as clock pulses come in, it won't be added to the total stored in the latch until the latch enable is brought low again.

Chip enable   The chip enable is on pin 11. Putting a high on this input means that incoming clock pulses won't be added to the total already stored in the chip's internal counter. A low on this input is the normal way the chip is configured because it enables the counter and allows the count to increase.

Reset input   The reset input on pin 13 is the clear control for the 4553's internal counter and latch. Making it high will store zeros in both places. A low on the reset input will allow the chip to operate normally.

Overflow output   The overflow output at pin 14 is used when two or more 4553s are put together to extend the counting capability of the circuit beyond the single-chip limit of 999. This pin puts out one clock pulse each time the count goes past 999. When you're cascading 4553s, it becomes the clock source for the next chip in line.

Oscillator inputs   The oscillator inputs on pins 3 and 4 are the place you put the timing capacitor for the chip's internal scan oscillator. This is the clock that strobes across the three digits in the display. The bigger the capacitor you use, the slower the oscillator will run. Typically, a value of 0.001 μF is used, but you can get as large as 10 times that without seeing any noticeable difference on the display.

Display outputs   The display outputs are $\overline{DS1}$ (pin 2), $\overline{DS2}$ (pin 1), and $\overline{DS3}$ (pin 15). These are the pins that connect to the LED cathodes and drive the displays. The chip strobes across these three pins in sequence so that only one display is actually being driven at any one time. Because the pins are active low, the 4553 is really only designed to drive common cathode displays.

Data outputs   The data outputs are on pins 9 (Q0), 7 (Q1), 6 (Q2), and 5 (Q3). These are the strobed output pins of the chip's internal latch. As the 4553 sequentially selects one of the three LEDs, it will put the corresponding data on the outputs.

Other pins   The other pins are power on pin 16 and ground on pin 8. Because the 4553 is a CMOS chip, it is extremely immune to noise and can operate safely on any voltage from about 3 to 15 V. As with any CMOS chip, however, the maximum operating frequency is dependent on the voltage at the power pin; thus, the 4553 has an upper clock limit of about 5 MHz when it's being fed with a 12-V supply.

The 4553 is a convenient chip to use because its internal counter can go up to 999, and it has all the logic built in to put the data out on three LED displays. At the moment, it's the cheapest multidigit counter/display chip around; and as far as I know, its primary source is Motorola and its secondary source is Goldstar.

Several mail-order houses carry the Motorola version, and you shouldn't have to spend much more than two or three bucks to get one. Because the sources for these chips are rather limited and your chances of picking one up anywhere in your neighborhood are about as good as finding a radiator ornament for a 1923 Bugatti, it's a smart thing to get a few of the 4553s when you put in your order.

Nothing is worse than blowing up the only chip you have and then having to wait who knows how long for a replacement. Remember that it's an eternal law of the universe that you'll smoke components when you're working on the breadboard. If you're really careful, even to the point of being tedious, you might be able to limit the destruction to only one chip but

The tendency of an IC to explode is directly proportional to the square of its cost and inversely proportional to the cube of its availability.

You can put this in symbolic form to make it look more scientific if you want; but however you express it, it's an unchangeable law—like gravity.

The 4553 and the 4511 are just about all you'll need to get a good display for the rate multiplier circuit we built together in the last chapter. I'm not going to break it down into a block diagram because the circuit is very simple. Even if you don't have much experience in electronics, you won't have any trouble getting it to work.

The schematic for the display is shown in Fig. 8-3, and you can see that we're not talking about a lot of complexity. Just about the only thing we haven't been through is the choice of display LEDs. The only consideration for the display LEDs is that they have to be a common cathode type. As long as you keep that one restriction in mind, you can use whatever display LED you want. As I mentioned when we were building the keyboard, I like FND500s because they fit on the breadboard, but you can use other ones if they're more to your liking. I understand yellow ones are very trendy now.

The display can be put on the breadboard as shown in the placement diagram of Fig. 8-4. Try to keep the leads as short as possible to make the board as neat as possible. I know it's a pain in the neck to cut and strip wire to exact length; but the neater the breadboard, the easier it is to troubleshoot and the less likely you are to accidentally pull wires out when you're working on it.

The only connections you have to make from the display to the rate multiplier circuit are from the 4089 multiplied rate output to the clock input of the 4553, and the reset pin of the 4040 to the reset pin of the 4553. You'll notice that we're not

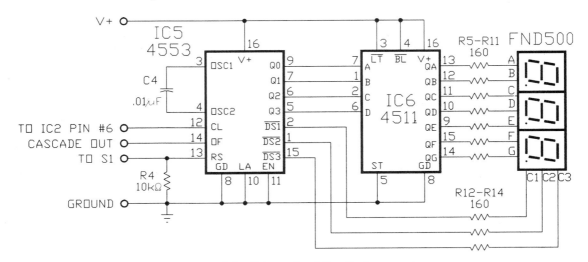

8-3 Schematic of the display.

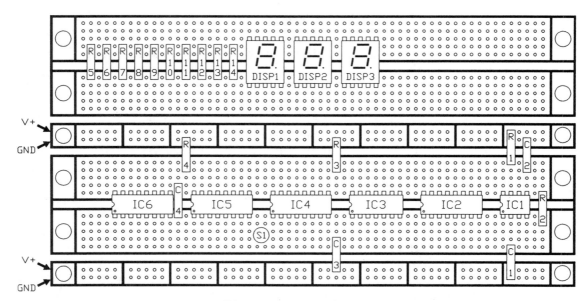

8-4 Placement diagram for the display.

using most of the 4553's control pins at the moment. The chip is kept permanently enabled and the latch is set to be transparent. Whatever pulses come into the IC are going to be counted and transferred, through the 4511, to the display.

We'll play around with the control pins later. At the moment, it's much better to set things up like this because you'll be able to see how the count increments when you press the reset button. If you slow down the 555 clock to 10 Hz or so, the display will be interesting to watch. You don't want to leave it like that in real-world applications, but it's a neat thing to see when you first get the circuit working. Besides, in most real applications, there's no need to see the result of the multiplication on a display because the real reason for doing the arithmetic is to use the resulting number in some other part of the circuit. However, it does look more impressive this way.

Even more arithmetic

What we've got working on the breadboard at the moment is a circuit that will take two predefined numbers, multiply them together, display the result on some LEDs, and at the press of a button do the same calculation over again.

The thrill doesn't last. Even if you rewired the circuit to change the numbers to your favorite ones, it won't be long before you start to get tired of the whole business. I can understand that. As soon as you reach this point, it's time to take the circuit a bit further. That means using more than one rate multiplier in the circuit. However, before we can start wiring stuff up, there's a bit more to learn because cascading rate multipliers together is a bit tricky.

One thing we can do at the moment is take a quick look at what else can be done with rate multipliers in general. We set things up to do multiplication, but regardless of what the chip is called, it can do virtually any arithmetic operation. The easiest way to see how this works is to modify our circuit to do some division.

The key to doing division with the rate multiplier circuit we have on the breadboard is to understand how it does multiplication. When we feed a clock to the chip, it gives us two basic outputs. The first is the base rate output, which (with the 4089) is one-sixteenth the clock; and the other is the multiplied rate output, which is the base rate multiplied by whatever number we happen to have presented to the chip's data inputs.

We can do multiplication because we get one base rate pulse out for each set of multiplied rate pulses. Multiplication, you'll remember, is just a matter of counting up the multiplied rate pulses for a given number of base rate pulses. All we're really doing to multiply two numbers together is successive addition.

Doing division isn't much different. Since we did successive addition to multiply two numbers, we'll do successive subtraction to divide two numbers. Here's how it's going to work.

In order to divide two numbers by a process of successive subtraction, we're really asking how many times we can subtract one number from another before we reach zero. After all, on its most basic level, that's what division really is. Keeping in mind how the rate multiplier works, we can get successive subtraction by doing the exact opposite of what we did for multiplication.

Instead of (as in multiplication) counting up the multiplied rate pulses for a given number of base rate pulses, what we do now is count up the base rate pulses that the chip outputs for a given number of multiplied rate pulses. The number of base rate pulses we get is the answer to the division problem.

The practical setup for doing this is something you should be able to work out for yourself; but to save you some potential brain damage, all it takes is one double-

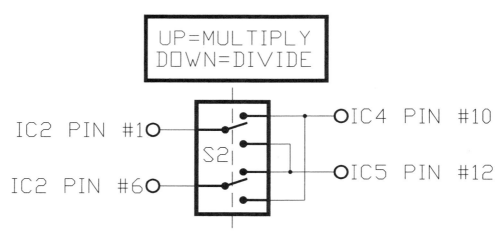

8-5 Using a switch to select multiplication or division.

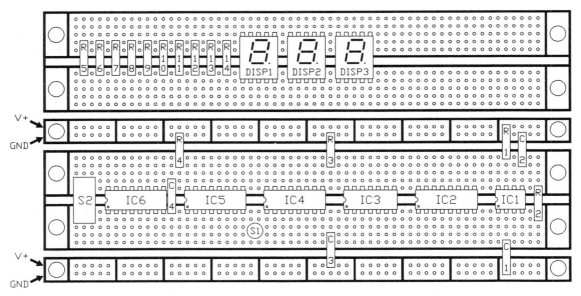

8-6 Placement diagram for the switch.

pole, double-throw (DPDT) switch. Believe it or not, that's the only thing you have to add to the board!

Since the only thing we want to change in the circuit we have so far is the destination of the 4089's base rate and multiplied rate outputs, all we have to do is a simple bit of rewiring as shown in the diagram of Fig. 8-5. Put the switch on the board as indicated in the placement diagram of Fig. 8-6, but don't try it yet. Remember that the circuit is hardwired to multiply two numbers—14 and 67—and switching it to do division means that you'll be dividing 14 into 67.

There's nothing wrong with doing this, but the answer you see on the display will show you one consequence of using a rate multiplier to do division. Throw the

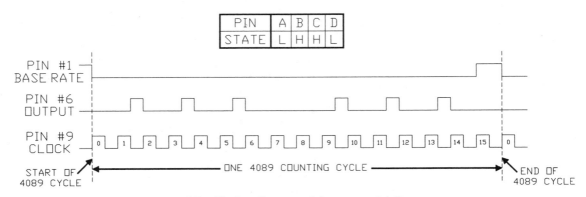

8-7 Timing diagram of the rate multiplier.

switch to the division position and press the reset button. You'll see 4 in the display instead of 4.79, the correct answer. The reason this happens is evident if you look at the rate multiplier's timing diagram (Fig. 8-7).

Notice that the base rate pulse doesn't get generated until the end of each rate multiplier cycle. There's nothing you can do to change this unless you get some tiny tools and do bypass surgery on the IC itself. This was common practice on Neptune for years, but it stopped when Motorola threatened to sue for patent violations. As the 4040 counts up the multiplied rate pulses from the 4089, four base rate pulses get sent to the display; but the 4040 reaches a count of 67 and stops everything in its tracks before the fifth pulse can be sent.

Rate multipliers always deal in whole numbers and round down when you use them to do division. The only thing you can do to increase your accuracy is to decide how many decimal places you'll want to have and plan for that number when you initially draw up your list of design criteria.

If you want an accuracy of two decimal places, you'll have to do a bit of prescaling before presenting the numbers to the circuit. Two decimal places means you'll have to multiply one of the numbers by 100 before the rate multiplier sees it. How you can do this depends on what you're doing to generate the two numbers in the first place. If nothing else suggests itself, you can always do the prescaling with a rate multiplier. Not bad.

Bigger numbers

While it may very well be true that the arithmetic you have to do in a circuit will always have at least one number that's less than 16, there's no guarantee that you won't have to do arithmetic with larger numbers sometime in the future. There are rate multipliers that have a wider data bus than the four-digit limit of the 4089, but sooner or later you're going to find it necessary to cascade two or more rate multipliers together.

Most counters and other chips that are designed to be cascaded have control pins on them to make the job as easy as possible. The rate multiplier is no different, but there are two basic ways it can be cascaded. The method you choose depends on the numbers you expect to see when you're doing arithmetic.

The standard way to cascade two or more rate multipliers together is shown in Fig. 8-8. The chip's designers call this method the *add mode*, and it's the most versatile of the two techniques. The number presented to the data inputs (one of the two numbers in your arithmetic problem) has to be broken into two halves. The first half goes to the first chip, and the second half goes to the second chip. Remember that each of the two rate multipliers can accept a number no larger than 16 at its data inputs. When you cascade chips to handle larger numbers, you have to give some thought as to how you're going to translate the number up into 8 bits.

As with just about everything in life, there are advantages and disadvantages. To get the bad news out of the way first, you can see in the schematic that the cascade input of the second chip is being driven by the multiplied rate output of the

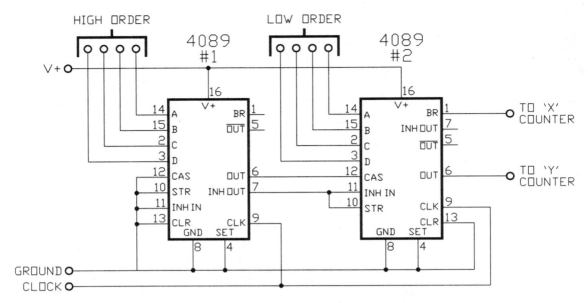

8-8 Cascading rate multipliers in the add mode.

first chip. The consequence is that the number you're using has to be broken into two parts in a way that takes into account the difference in the output rates of the two 4089s.

Let's go through an example. If we're doing arithmetic and one of the two numbers we're using is, say, one hundred—might as well make this example as useful as we possibly can—we have to break the number into two parts that, when added together, result in a total of 100. This would seem to be a simple operation, but remember how the rate multiplier works. We get an answer by using both the base rate and the multiplied rate. Since the cascade input of the second rate multiplier is being clocked by the multiplied rate output of the first one, you're going to get different pulse rates from the multiplied rate outputs on each chip.

When we divide the number we're using—in this case 100—into two 4-bit sections, the most significant 4 bits will go to the first rate multiplier and the least significant 4 bits will go to the second rate multiplier. The connection between the inhibit output of the first rate multiplier and the inhibit input of the second causes the multiplied rate pulses from chip 2 to be at one-sixteenth the rate of the ones coming from chip 1.

Since the multiplied rate pulses we'll be feeding to the external counter come from the second rate multiplier, and there's a factor of 16 between the two chips, we have to divide our number, taking this difference into account. This means we have to reduce both the most significant and least significant parts of the number to the lowest common denominator. The first rate multiplier is working with a base of 16, but the second one is working with one-sixteenth of that, or 256.

In order to have the two rate multipliers deal with a number like 100, we have to break the 100 up into two numbers, A and B, that will satisfy the equation

$$100/256 = A/16 + (B/16)/16$$

or $\qquad 100/256 = A/16 + B/256$

The reason we add them is that the first 4089 controls the enabling of the second 4089 through the connection made between the inhibit output of the first chip and the inhibit input of the second. The result of this connection is that the two chips behave as if they were in series, with the first counting decades and the second counting the remainder.

To divide 100, we divide it by 16 and put the integer portion on the data inputs of the first rate multiplier and the remainder on the data inputs of the second rate multiplier. Simple arithmetic gives us a value of 6 for A and 4 for B.

Now keep your eyes open for this one. For a chance to win a three-week (20 sun-filled days and one fun-filled night) vacation on Neptune (transportation not included) and a lifetime supply of hula hoop lubricant, who can tell me what the relationship is between 100 (the number we started with) and 64 (the number presented to the rate multiplier)?

That's right—100 decimal is the same as 64 hex.

Remember that the 4089 is a *binary* rate multiplier; and because it's working with a base of 16, the numbers it wants to deal with are binary ones usually expressed in hex. The answers you get, in terms of the number of base rate and multiplied rate pulses, can be counted using either number system—after all, a pulse is a pulse is a pulse.

The bad news about cascading 4089s with the add mode is that it takes some thought to convert the number to hex. The good news is that you can avoid the mental strain and substitute the 4527 *decimal*-based rate multiplier for the 4089 and deal with base-10 BCD numbers instead of the base-16 straight binary numbers required with the 4089.

And yes, the pinouts are the same so you can pop the 4089s out of their sockets and replace them with 4527s.

Of course, using 4527s means you'd need three chips to handle the number 100 because the BCD inputs can only deal with a maximum count of 9. If we express the data input requirements for the BCD version of the rate multiplier in the same format we did earlier for the 4089, you'll see what I mean.

Just as before, we have to break the number 100 into, in this case, three numbers, A, B, and C, that satisfy the equation

$$100/1\,000 = A/10 + (B/10)/10 + (C/10)/100 \,.$$

Since the decimal chip is working with a base of 10, we have to have a third place for the hundreds. As shown above, A counts the hundreds, B counts the tens, and C counts the units; and we wind up with

$$100/1\,000 = 1/10 + 0/100 + 0/1\,000.$$

While using decimal-based rate multipliers means you don't have to do any decimal-to-hex number conversion, you'll more than likely wind up using more ICs. Chips are cheap but the more of them you have in a circuit, the more complex the board is going to be and the more likely you are to make a mistake.

The add mode we've been discussing can accommodate any number at all. However, there's another way to cascade rate multipliers known as the *multiply mode*. This mode, while requiring fewer connections between the chips, is more limited in the numbers it can handle. (Refer to Fig. 8-9.)

When the rate multipliers are set up like this, you still have to break the number at the data inputs into two parts; but because the multiplied rate of the first chip is feeding the clock input of the second, the 4-bit parts you have to create will multiply rather than add. This can be seen more clearly when we look at it in the form of an equation.

If we wanted to do arithmetic with 100 at the data inputs of two binary rate multipliers, we'd have to break the number into two parts so that

$$100/256 = (A/16) \times (B/16).$$

Since we're limited by the fact that 16 is the largest number we can have at the data inputs of either chip, the only two numbers that can work are 10 and 10.

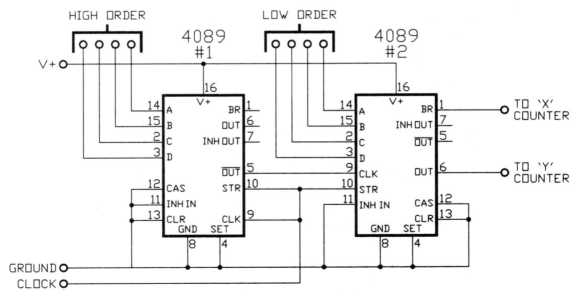

8-9 Cascading rate multipliers in the multiply mode.

There's nothing wrong with doing this; but if you have a number that can't conveniently be divided into two whole numbers, you're out of luck.

This circuit arrangement, the multiply mode, is a convenient way to build a circuit that can square numbers; but a trip to a math book will show you that there are ways to do squares, logarithms, and even cube roots with nothing more than simple multiplication and division. The bottom line is that while the multiply mode is easier to wire, the add mode is more versatile and much better suited to being fed by a standard data bus.

No matter which rate multipliers you use, it should be clear to you by now that these aren't really the kind of chips that are made to be used with rotary switches and other mechanical devices. The natural home for a rate multiplier is in the middle of a populated circuit, and the data should come to it from other parts on the board.

Doing real work

Don't fall into the trap of thinking that rate multipliers are hard to use and need a lot of special circuit considerations before they'll reliably do what they were intended to do. Your mind may have been a bit boggled by the last couple of pages, but it's not really necessary to understand all the grimy details of rate multipliers to be able to use them in a circuit.

The time has come to do some work on the demonstration circuit you have sitting on the board. It may be able to multiply and divide two numbers; but the limitations caused by both the mechanical switches and gated decoders make the circuit, if the truth be known, not much more useful than single-ply toilet paper.

The biggest limitation of the circuit at the moment is the 4040 we have for the Y counter. While long-chain counters are a quick-and-dirty way to get big counts in a single package, the only thing they can do is count up. This is a problem because the output decoding arrangement has to change for each number you want to detect, making it all but impossible to feed in a number from a bus.

The only way we can do this is to have a counter arrangement that can be loaded by simply putting a number on a bus and turning the chip loose. The answer is that rather than starting at zero and counting up to a particular number, we'll start the count at that number, count down to zero, and stop the rate multiplier when zero is detected. By doing this, we can lock in our gating setup and make the circuit work for us no matter what numbers we want to use. The only restriction will be the width of the buses.

Most of the requirements for this circuit were spelled out in the last chapter when we drew up the list of design criteria to cover the demonstration circuit. In order to convert that circuit into a more general one we can use in real-world applications, we only have to add two more criteria to the list.

1. The circuit will be able to do arithmetic with two 8-bit bytes.

While we're free to make the bus as wide (or as narrow) as we want, most applications will want to have a full 8 bits available for data. You don't have to use all of them. If you're some atavistic octal throwback, feel free to ignore whatever bits you want.

2. The circuit will have a full set of control lines.

We've been starting the circuit operation with a mechanical switch, but we want this circuit to be a complete subsection ready to be dropped into the middle of another application. There's no reason why we can't leave some parallel mechanical controls; but if there aren't control lines as well, the circuit's not going to be too useful.

We've been through this together before so you should know by now that the next step in the design process is to lay out the block diagram for the circuit. This isn't going to take too much brain strain or energy because, as shown in Fig. 8-10, the block diagram for the demonstration circuit is pretty close to what we want to build.

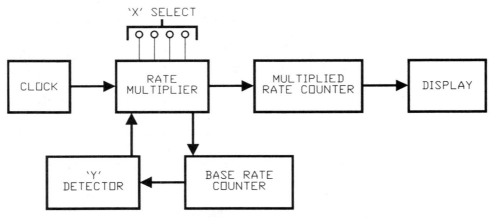

8-10   Block diagram of the demonstration circuit.

The additions we're planning to add are reflected in the updated version of the block diagram (Fig. 8-11). I've widened the bus to 8 bits, replaced the original Y detector with a down counter and zero detector, and (something new) indicated that we're going to be needing a control circuit as well. It may look a bit more chaotic than the earlier one, but don't forget that

There is no such thing as chaos, merely unperceived order.

So let's get down to making it all happen.

One of the inescapable truths in the world of electronic design is that you can't design a control circuit until you know what you want to control. Doing arithmetic is going to be a matter of putting the correct bytes on the correct buses, throwing the right electronic switches, and finally turning the circuit on to do its very own electronic thing. Sounds simple enough.

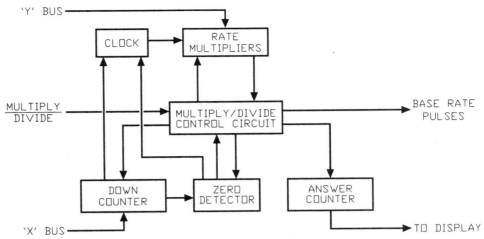

8-11 Block diagram of the final circuit.

Because a couple of new things are indicated in the block diagram and we've got to take care of them before we reach the end of this chapter, we might as well start on the new Y detector. The detector has three basic requirements:

- It has to be able to be preloaded with a particular number.
- It has to be able to count down from that number to zero.
- It has to be cascadable.

Anytime you find yourself drawing up a list like this, it can only mean you're in for a trip to the magic world of databookland. Now I can only guess that yours haven't come in the mail yet, but among your collection should be good catalogs of basic MSI logic stuff—both TTL and CMOS. It doesn't matter what you design— from mundane stuff like electric shoelaces to super exotic stuff like bionic radar (available in kit form from Zorch Electronics, $200 FOB Neptune), you're always going to need counters, gates, and all the other basic goodies listed in these books.

A good choice for our new Y detector is the 4516, whose pinouts are shown in Fig. 8-12. This binary up/down counter can be preset, is cascadable, and has a BCD clone in the 4510.

You know what comes next.

Parallel jam inputs   The parallel jam inputs are weighted in binary and, as is usually the case, located in a typically random fashion. P0 is on pin 4, P1 is on pin 12, P2 is on pin 13, and P3 is on pin 3. These are the inputs that are used to load whatever number you want to use for the initial count.

Data outputs   The data outputs are also binary and are Q0 (pin 6), Q1 (pin 11), Q2 (pin 14), and Q3 (pin 2). When the chip is operating, this is where you get the output from the counter.

Load input   The load input on pin 1 is the control pin that's used to preload the counter. By putting a number on the four jam inputs and then hitting this input with a positive-going pulse, you'll load the data into the counter and that will be the

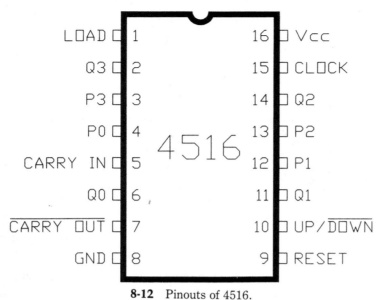

8-12 Pinouts of 4516.

starting number for the count. The clock input has to be low when you preload data.

Clock input The clock input on pin 15 is where you feed in the clock the chip uses for counting. The count is done on the positive-going edge of the clock pulses. As with all CMOS parts, you want to feed this input with a suitable clock— one with square shoulders and fast rise times.

Up/down input The up/down input on pin 10 controls the direction the counter will take. When a high is put here, the counter will add. If it's low, the counter will subtract.

Reset input The reset input on pin 9 will clear the counter to zero when you make it high. Since the counter can be preloaded, you can also clear the counter by jamming in a zero but this is a bit of overkill—kind of like using heavy machinery to lace up sneakers.

Cascade input The cascade input is on pin 5 and is used when two or more chips are tied together. This pin is tied low when only one chip is used; and when chips are cascaded, the first one in line should also have this pin grounded.

Cascade output The cascade output on pin 7 feeds the cascade input of the next chip in line when two or more chips are cascaded. It's active low and will output a pulse when either a zero or 15 count is reached in the counter.

Other pins The other pins, finally, are the power and ground connections at the standard locations of pins 16 and 8, respectively.

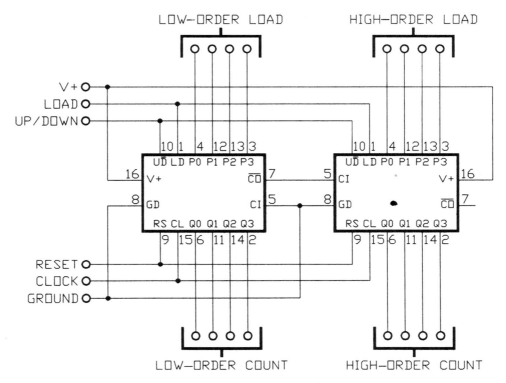

8-13   How to cascade 4516s.

Because we need an 8-bit bus, two of the 4516s have to be cascaded (see Fig. 8-13). Both chips are being clocked by the same source, the cascade output of the first chip is connected to the cascade input of the second, and both load inputs have been tied together so the 8-bit data can be loaded into both counters at the same time. I've also connected both of the reset pins because whatever happens to one counter has to happen to the other one as well.

One important thing we have to fix in the demonstration circuit is the way we choose between multiplication and division. It may be OK for a demonstration circuit to have a mechanical switch, but even the world's smallest microswitch with a throw pressure that's measured in micronanograms will probably still be a bit too stiff to be thrown by electrons. The mechanical switch has to be replaced with something else that can be handled by electrons.

The perfect choice is a 4066 analog switch whose pinouts are shown in Fig. 8-14. This is nothing more than four independent single-pole, single-throw switches in a single package. Each switch has an input, output, and control pin. A high on the control pin closes the switch and a low opens it. Since this is an electronic device, you don't get the zero "on" resistance and infinite "off" resistance found on its mechanical equivalent, but you'll get pretty close to that.

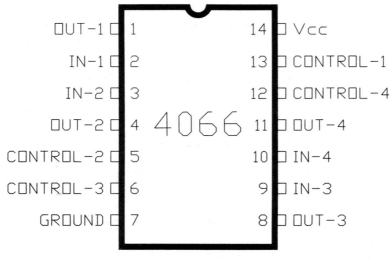

8-14   Pinouts of the 4066.

The on resistance is about 75 Ω and the off resistance, while not the open-air resistance of a mechanical switch, still has about 10 zeros on the left side of the decimal point. If you ever find the on resistance to be too high for your application, you can always put a couple of the switches in parallel to drop the resistance down to a number you can be happier with.

Analog switches like the 4066 can be configured to duplicate any of the standard mechanical switches. As shown in Fig. 8-15, the chip is wired as a DPDT switch. Compare this illustration to Fig. 8-16, the mechanical switch setup we used in the demonstration circuit.

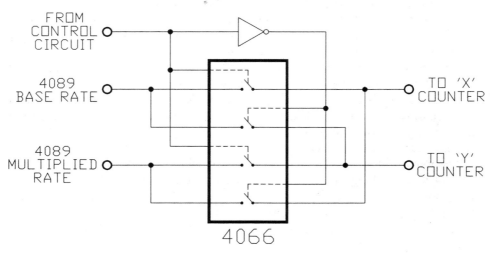

8-15   Configuring the 4066 as a DPDT switch.

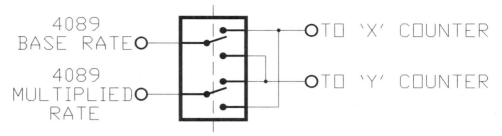

8-16   Mechanical equivalent of 4066 configuration.

By putting the inverter in the switch circuit, we're accomplishing two things. First, the inverter guarantees that we'll never be able to have an illegal state on the switch. Second, we only need one line to control the switch. Even though I used an inverter, you can replace the inverter with any device that will take an input signal and turn it upside down—anything from a single transistor to any spare inverting gate you have lying around the circuit.

The whole rate multiplier circuit is designed to be like a drop-in module that can be dumped into the middle of any project that needs to have some arithmetic done—think of it as your basic black box. If we did put it in a black box and just brought the data, control, and power lines out to the rest of the world, you could treat the circuit as if it were just an oversized integrated circuit.

In order for this arithmetic module of ours to be able to work as an independent module, it has to be designed so that it can be driven in exactly the same way any IC is driven—completely untouched by human hands. The control lines receive operating instructions from the outside world; and once that's set up, the rate multiplier circuit is configured to do the desired job and turns itself on to get the job done.

One really worthwhile goal to have in mind when you're designing a subcircuit like this one is that the circuit's operation should be as automatic as possible. There should be a minimum of lines fed to the outside world because the more signals you have to deal with, the more likely it's going to be for you (or whoever uses the circuit) to screw it up. Not only that, but lots of lines will be translated into lots of traces on a PC board and, believe me, you really want to keep those to the absolute bare minimum.

In the best of all possible worlds, this rate multiplier circuit should only have the following lines going out to the rest of the world:

1. 16 data input lines—8 each for the X and Y numbers
2. 16 data output lines—the multiplied maximum
3. A pulsed output line—a serial version of the data output
4. An enable control line—to start circuit operation
5. A completion signal—to signal the end of the calculation
6. A multiply/divide control line—for the 4066
7. Power and ground

The second item on the list may have thrown you. Remember that with eight X and eight Y lines, the biggest multiplication problem we can do is FFh times FFh or, in decimal, 255 times 255. This means we could conceivably wind up with an answer as large as 65 025 or, in hex, FE01h. Since each one of those hex digits is really 4 bits, we might need as many as 16 data output lines to get the answer out to the rest of the world.

There aren't any convenient counters that can output every number from zero to 65 025 on 16 data lines, but we can get around that by using two counters cascaded together. Since the only requirements we need in the counters are a clock and a reset, there's no shortage of possible candidates.

We've already spent time with the 4040 so we might as well use two of them. Even though there are no convenient cascade controls on these chips, all we have to do is feed the clock input of the second chip with the last output of the first chip. Of course, we also have to tie the two reset pins together because, just as with the 4516s, whatever happens to one chip has to happen to the other one as well.

The 4040s will give us the answer to the arithmetic problem in binary form, but it's not so unreasonable to expect that there might come a time when it would be better to have the circuit just give us a train of pulses that can be counted by something else. This could be important if you wanted to know about an interim point in the count or if you were controlling a stepper motor and the calculation result was being used to move the motor a certain number of steps. In any event, because it's so easy to generate this kind of output, there's no reason not to do it.

Making it happen

We've dissected every piece of the circuit we need and gone through the modifications we have to make to the demonstration circuit so all that's left is to put the whole thing together. The schematic for the complete circuit is shown in Fig. 8-17. While most of it is just a clone of the demonstration circuit (the schematic in Fig. 8-18), you'll notice a few extra changes that we haven't talked about yet.

In order to make the operation of the circuit completely automatic, we have to find a signal that indicates the calculation is completed and use that signal to stop the circuit from doing any further counting or clocking. Since the arithmetic operation is finished as soon as the count in the 4516s reaches zero, the output of the zero detector is the signal we're looking for. When this signal is generated, we want it to make the rest of the circuit freeze in its tracks, as it was doing before in the demonstration circuit. We also want to make this signal available to the outside world as the completion signal we listed earlier.

There are lots of counters and other things in this circuit that have to be disabled for us to stop the circuit dead in its tracks at the end of the calculation. I've summed it all up in Fig. 8-19, and you can see that we're looking at a collection of highs and lows. The output of the zero detector is the one signal that has to do all this. Although it's no big deal to invert the signal so we have both a low and a high trigger available, we need an inverter to do the job.

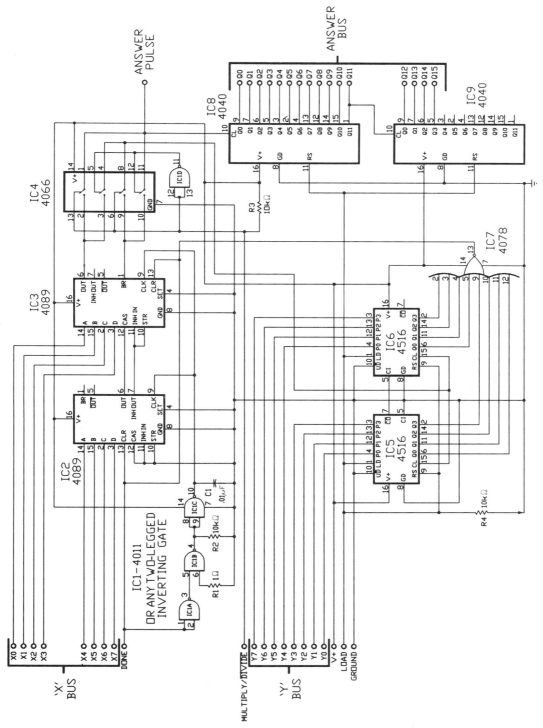

8-17 Schematic of the final rate multiplier circuit.

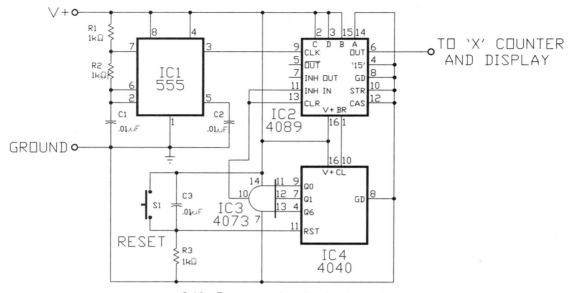

8-18 Demonstration circuit schematic.

As we've been developing this circuit, we've been adding inverters to the board. We now need one to invert the zero detect line, and we also need one to build the double-pole, double-throw switch with the 4066. Since we have to put some on the board, and since it's always a good idea to keep the total parts count as low as possible, you'll notice that I've replaced the 555 clock we had in the demonstration circuit with a gated oscillator made out of two inverting gates. The two NAND gates left in the package have had their inputs tied together (and if that

4089 CIRCUIT CONTROL LINES			
PART	PIN	SIGNAL	RESULT
IC1 - 4011	#1	LOW	STOP
IC2 - 4089	#13	HIGH	STOP
IC3 - 4089	#13	HIGH	STOP
IC5 - 4516	#9	HIGH	RESET
IC6 - 4516	#9	HIGH	RESET
IC8 - 4040	#11	HIGH	RESET
IC9 - 4040	#11	HIGH	RESET

8-19 Control lines of the rate multiplier circuit.

doesn't sound awful, nothing does) to turn them into a pair of simple inverters. That means no wasted silicon—a good thing.

The operation of the circuit is controlled by the enable line we've brought out to the real world. Once the data has been presented to the 4089 and 4516 buses and the multiply/divide control has been set, the rate multiplier circuit is started by putting a brief positive pulse on the enable line. This action resets the 4040s and causes the number at the jam inputs of the 4516s to be loaded into the chips' internal counters. As a result the output of the zero detector goes low which, in turn, enables the 4089s and, through the inverter, puts a high on the enable pin of the clock.

The clock starts putting out pulses, and the circuit begins doing the arithmetic calculation. When the count in the 4516s reaches zero, the calculation is finished. The zero detector goes high and disables the clock and the 4089s. The answer remains available at the output bus of the 4040s until the external enable signal is triggered again to mark the start of another calculation.

I've put resistors on the circuit's two control lines just to be sure that they won't ever be left floating—a big no-no when you're using CMOS stuff and an equally bad habit with any other logic family.

The operation of the circuit is, as any circuit of this kind should be, completely automatic. It's not entirely bulletproof, though, because there are some things that can screw up the way the circuit operates. I'm not going to tell you what they are, by the way. You'll find out for yourself. Trust me.

No circuit in the world is completely bulletproof—any more than a piece of software is completely bulletproof. You just can't design an absolutely foolproof anything. It's a law of the universe that an end user is the most creatively destructive person in the world. It may be unintentional, but no matter how much brain damage you go through to tie up all the loose ends in a design, end users will find one you overlooked.

The only way to be sure that something can't be screwed up whenever you're not around is to limit the options you make available to the user. The circuit we've been working on only has a few lines that have to be controlled in order for it to do arithmetic. You can add as many other options as you want but, as a general rule, less is better.

You can get the circuit working by laying it out on the breadboard as shown in Fig. 8-20. As always, keep the leads as short as you can and, if possible, clip the component leads to keep them short as well. There's nothing critical about the design; and just about the only thing you might find annoying is that the LED display is still limited to three digits.

As I said before, the display isn't an integral part of the circuit; but if you want to make it wide enough to display five digits (the largest answer you can get), you'll need another 4553, another 4511, and some more LED displays. The schematic in Fig. 8-21 shows what has to be done to cascade the 4553s. It's so simple to do, you might want to wire it into the breadboard just to have a convenient way of checking the circuit.

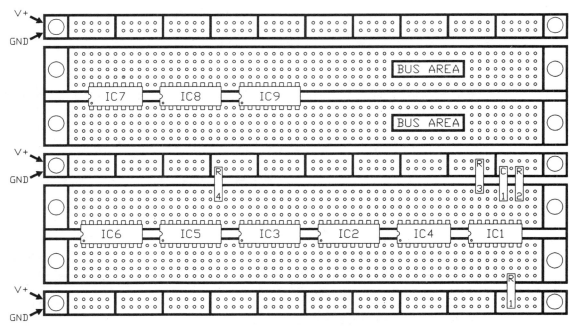

8-20 Placement diagram for the rate multiplier circuit.

If you do decide to add the display, put it on the breadboard as shown in Fig. 8-22.

Oops!

There's enough of a difference between the demonstration circuit and the final one to cause screwups in assembly. A bunch of stuff has to be removed from the board and new stuff added so it's always possible to pull something loose or put a wire in the wrong spot.

As with everything else we've done, the best way to troubleshoot this circuit is to isolate each section and then check to see which sections are working and which are dead. Just start at the beginning and work your way to the end. Ask yourself basic questions like is the clock clocking, are the counters counting, are the data lines moving up and down—stuff like that.

Make sure you have power and ground every place they're supposed to be and nowhere they shouldn't be.

No matter what part of the circuit you're having trouble with, the order of things to suspect is always bad connections first, then wrong component values, then dead chips. The chances of getting dead ICs from a supplier are really small; and unless you've inadvertently blown them up yourself, chip substitution is something you only do as an absolute last resort.

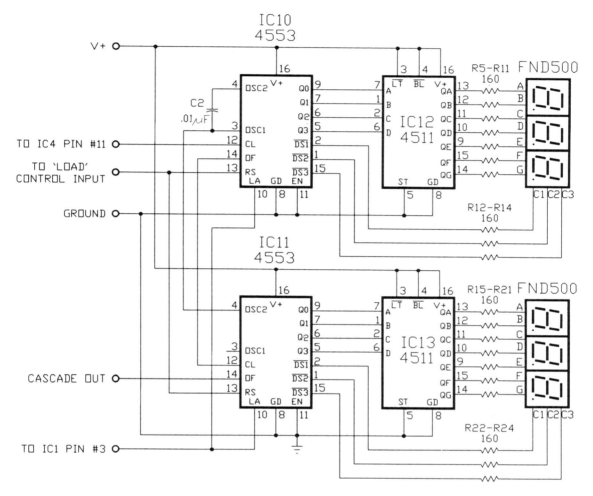

8-21 Cascading 4511s and 4553s to get a six-digit counter.

Pulling chips from a breadboard is almost always guaranteed to move wires around (so they pull loose) or move components around (so they touch). Whenever you substitute chips, take some time to recheck the surrounding wires and components before you repower the board.

You did remember to pull power, didn't you?

So now what?

Having a working circuit like this is a bit of a bummer because it's not the kind of thing that does anything on its own. What you really have in front of you is something that's made to be dropped inside a circuit that needs some arithmetic done.

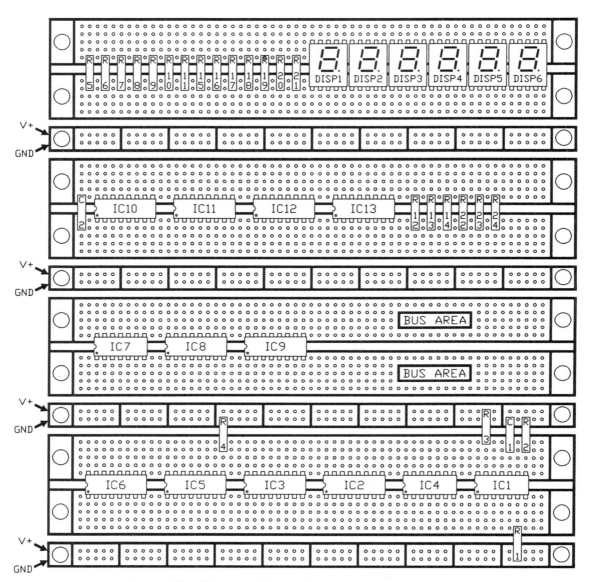

8-22   Placement diagram for the display circuit.

It is kind of neat to be able to manually put numbers on the bus, set the multiply/divide control, push the button, and have the result appear on the display. The only problem with doing this is that you need some method of entering the numbers from a convenient keypad and putting them on a bus . . . but hold on a minute, haven't we run across something like that? It sounds familiar.

That's right. The keyboard circuit we just finished can feed this one without any extra brain damage whatsoever. All we'll be using are the lower four digits of

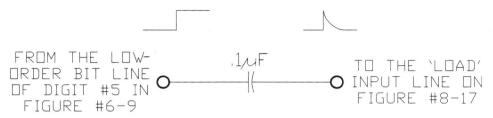

8-23 Using an RC combination to get a pulse from a steady line.

the keyboard's output bus (two for the 4089 and two for the 4516). We'll still need a momentary switch to generate the load pulse, but that's trivial.

You can build yourself a half monostable or just take one of the bus lines associated with the fifth digit. If you feed the load input with, say, the low-order bit of the fifth digit, you'll generate a high by entering the fifth digit as a 1, 3, 5, or any number that makes the low-order bit go high. Since you want a pulse rather than a steady high, put a capacitor on the line as shown in Fig. 8-23. This will generate a fairly sloppy pulse, but it's more than adequate for the load input.

You can use the same technique for the multiply/divide switch. All you have to do is pick a convenient bit on the keyboard's external bus and connect it directly to the multiply/divide control line of the rate multiplier circuit. You don't want a capacitor here because this line wants to see a constant high or low, not a pulse.

Don't forget that both circuits need a common ground.

Being able to do electronic arithmetic is something that can often stump even a veteran designer. In these days of microprocessor-based everything, lots of circuit designers have gotten in the habit of just dropping a microprocessor into a circuit and then using a few words of software to do whatever arithmetic has to be done. This makes things more complex, more expensive, and—worse yet—a lot less reliable.

Rate multipliers are one of the family of chips that are frequently overlooked, and why this is true is something I've never been able to understand. They're easy to use and can be an extremely simple answer to what might otherwise be a mind-boggling circuit problem.

But enough about rate multipliers. You now know more than most designers about how to use these chips. Feel free to go out there and experiment with them. The circuit we have working is a terrific addition to any designer's notebook, and it's also one that can save you a load of brain damage when the need for arithmetic arises.

All we've done with the circuit is multiplication and division, but that's all you really need because you can do virtually any arithmetic operation with just these two tools. Any good math book will tell you what the techniques are for taking roots by doing successive division. Squares can be done as well with successive addition. The best part of all this is that once you understand the method, the required circuit is almost exactly the same as the one we just completed. All it needs are a few additional gates. But I'll leave that to you.

Give yourself a couple of well-deserved pats on the back, take some time off, and when you get back, I'll be waiting here with something completely different. Count on it.

A slick trick

You can never have enough clocks in a circuit because many components need them to operate. If your application needs precise and stable frequencies, you have to be careful about how you build the clock. If you just need a train of pulses to drive a circuit that has to scan at high frequencies, this is by far the simplest way it can be done.

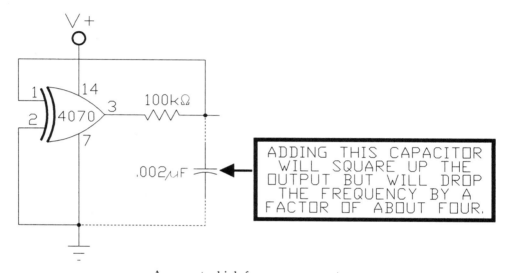

A one-gate, high-frequency generator.

The frequency from the circuit is going to depend on the supply voltage, resistor value, and propagation delay of the gate, but it will always be up there in the megahertz range. A 10-kΩ resistor and 5-V supply will give you a frequency of about 5 MHz.

9
Remote control #1
Getting there without leaving here

It's confession time. One of the main reasons I got interested in electronics comes from something that happened when I was about nine years old. I know you didn't buy a book on electronics to find out about my childhood, but I'm writing this thing. You can skip this part if you want, but if you do, you'll never know exactly what happened to me. And remember, knowledge is power.

When I was nine years old, my father bought a new car. The minute he brought it home, he insisted that we go for a ride. This was the first new car we ever had and I can still remember how the seats felt, how the interior smelled, how clean the windows were. I remember all the shiny knobs on the dashboard. You know, I've owned lots of new cars since my father brought that one home but none of them ever felt as good as that one did.

In any event, my father was as much of a gadget freak then as I am now, so the car was loaded with every conceivable accessory offered by the manufacturer. But what I remember most was the radio.

He turned it on as we were riding down the street; and every time he pointed his finger at it, the station would change. I could see that his hands were empty and he wasn't going anywhere near the radio; but no matter how closely I watched, as soon as he gestured at the radio, the station would change.

When I was nine years old, I wasn't as skeptical as I am now—close to it, though—and, after all, this was my father telling me it was magic so I didn't know what to think.

If he had let it go there, he would have had me, but then my father blew it. He pointed with his elbow and the station changed. He raised his eyebrows and the station changed. He looked at the radio and the station changed.

I might have been able to swallow all that but when he started with magic words like *abracadabra*, I knew there was something else going on.

I hate keeping you all in suspense (or putting you all to sleep), so I'll tell you that what was going on was simply that there was a button on the floor that my father would step on to make the station change. This may seem to be anticlimactic but it was a different time back then—and to a nine-year-old, this was a big thing.

We started talking about how it worked, how a button in one place could make something happen somewhere else, and then on to a nine-year-old's fantastic ideas about space travel, death rays, and (my personal specialty) button pushing.

From that moment on I was interested in electronics and started taking the backs off radios, getting shocks from capacitors, finding out (painfully) about filament transformers, and trying to understand why the radio's dial light wouldn't come on as soon as you turned on the radio.

OK, you can wake up now.

Everybody out there is an electronics junky, but electronics is a big subject. Every day that goes by brings even more areas into the world of electronics. Microprocessors and miniaturization have opened whole new worlds to electronics. Breakthroughs in imaging and sensor technology have resulted in the development of what have to be called revolutionary techniques and devices to learn about everything from the inner man to the outer planets.

The *Pioneer* and *Voyager* probes sent back incredibly detailed images of the solar system, but what most people don't realize is that both of these probes worked and transmitted with less than a hundred watts of power. It takes almost 10 times that amount to start your car on a cold morning!

The bottom line here is that as more and more territory gets taken in by electronics, the more likely it is that you'll be able to find a subject that interests you. This is where the hobbyist has a really big advantage over the professional designer. If you don't use the breadboard to pay the rent, there's nothing stopping you from doing only the things that interest you. A professional designer has to go for the green, but hobbyists can go wherever they want.

Even though any design project can take over your life, the only ones that grab 135% of your interest from the very beginning are the ones that are in areas you have an interest in.

For me, that's remote control.

The basics

Ever since that day in the car with my father, I've been fascinated with being able to push a small button here and make a big something happen there. Now remote control is a big subject. It covers just about everything from whistle switches to space probes and can use any transmission method from ultrasonic sound to lasers. Each of these has its advantages and disadvantages, and it's the designer's job to pick the best method for a particular application.

For a hobbyist, the choices are usually somewhat limited. Some of the transmission methods have to be dismissed because the technology uses parts that are all but impossible to get, unbelievably expensive, and dangerous to handle. But

9-1 Block diagram of a typical remote control system.

there are several methods that can be used without having to worry about price or availability; and once you get a handle on one of them, you'll be able to deal with any other one.

On a certain level, all remote control systems are the same. This may sound a bit strange; but as you can see in Fig. 9-1 (the block diagram for a typical setup), if you reduce everything to the bare-bones minimum, most of the basic elements are the same. Information gets encoded in the front end of the transmitter. This is usually a keyboard setup that produces unique keycodes for each available button.

These codes are sent to the second part of the transmitter, which is usually a completely independent circuit. This section of the system is the part that prepares the signals so they can be sent out to the receiver. While the keyboard circuits for remote control systems are essentially the same, the details of the modulator depend completely on the transmission method. Some of them, as in a car burglar alarm, are small enough to hang off a keychain, while others, as in the case of *Voyager*, fill a large building and need as much power as the island of Bora Bora.

The choice of transmission, as I mentioned, is always a practical one. You have to weigh your needs against the design hassles and the cost. A good rule of thumb here is that the farther apart you want to keep the transmitter and receiver, the more bucks you'll have to shell out—and if you're building everything yourself, the more brain damage you're going to come up against.

The choice of methods depends on the application. Using klystron tubes to get a signal across the room may be an interesting exercise in design, but you'll probably wind up taking the paint off the walls as well. By the same token, trying to go a few miles with infrared is pretty silly. Unless you use lots of batteries and lenses, the transmitter's signal is going to be just about as weak as a diplomatic protest.

There's no handy rule to follow about the required strength of the transmitted signal. In general, it's always better to make it a bit stronger, especially if you can't guarantee that you'll always have the same conditions in the space the signal has to cross. Having too weak a signal can result in flaky operation, and there's not much sense in designing a system that leaves you constantly worrying if you can get the signal to the other side of the room, much less out to the void.

The receiver in a remote control system is the mirror image of the transmitter. The demodulator converts the transmitted signal back to its original form and presents it to the circuitry that understands what each keycode is supposed to do. In case you're wondering, there's no reason why you couldn't take the signal from the transmitter's encoder section and connect it directly to the decoder in the receiver. As a matter of fact, when the system is still under development, you should always do that before the transmission section is designed.

Remember that the point of block diagrams and modular design is to let you verify each component of a circuit separately. The more complex the project, the more important it is to know that each section works correctly before putting them all together and trying the system as a whole.

Planning a system

After all we've been through so far, it better not be a surprise that the first step in the project is drawing up a list of criteria. I know that I've been repeating this throughout the whole book, but I'm not doing it to make things tedious. I'm repeating it because it's one of the most important parts of the project. Without a good idea of what you want to do, you won't be able to do a good anything. Guaranteed.

Even though the remote control system we're designing will be more complex than most of the other stuff we've done, the list of criteria will be shorter. This may seem strange but the reason is simple. While the project may be complex, the job it has to do is pretty basic. The goal we're aiming for is just to produce data at the front end of the transmitter and have the same data appear at the back end of the receiver. If we weren't going to need a receiver and transmitter, all we'd be looking at would be a simple keyboard encoder, and we've already seen that it doesn't take much to put one of those together.

If you look over the keyboard circuit we designed several chapters ago, you'll see that while the complete circuit is complicated, the basic keyboard encoder part of it is simple. The majority of the circuitry we designed back there was for the purpose of building a multidigit data bus, not generating the keycodes themselves.

Since we're going to be designing a remote control system, we need more than a basic encoder/decoder circuit. We also have to work out the transmitter and receiver as well as the circuitry to modulate and demodulate the transmitted signal.

But let's draw up the list and see exactly what has to be included on it.

1. The system will be able to issue at least 16 commands.

The problem with designing something without any real application in mind is that it's hard to make clear-cut decisions. Because there's no definite job the circuit has to do, there's nothing to serve you as a guide for the best way to configure everything.

We've already discussed the fact that keyboard encoder circuits can be designed several different ways. When we did ours, we used two or three chips to get the job done. We'll need another keyboard encoder here and, while there's nothing wrong with using the one we did before, this is a good opportunity to stretch our brains and do something different.

The keyboard encoder for the remote control system will use the same basic idea as the one we did earlier, but we'll design it differently. Because the earlier one was decimal based (10 digits), we might as well make this one hex based (16 digits). Both of these are expandable, but we never looked into that before so we'll do it here. Remember, it's only the technique that's important—not the particular components. Anything we do here can be done back there.

2. The remote control system will use infrared light.

I thought about this one for a long time. There are slicker ways to get a signal across a room, but there aren't any that are as reliable and cheap. Ultrasonics makes my teeth hurt, drives my neighbor's dog nuts, and my friend swears he can hear it no matter what frequency I pick. Lasers are neat but you can't see the infrared ones; HeNe's and other gas discharge lasers are sexy but batteries don't last too long; laser diodes are just too expensive; and (bottom line) I have really lousy aim.

While it's true that infrared has a somewhat limited range, most remote control systems never have to go more than about 30 feet or so; and if you absolutely have to, you can up the output power of the transmitter and design the receiver to be more sensitive. All modern remote control systems for consumer electronics use infrared because it's relatively easy to design around and the power requirements are very low. Even though you're using infrared LEDs, they're a lot more efficient than visible LEDs and give you more power at the output for the same number of amps at the input.

One other advantage of the infrared system is that it's capable of transmitting analog as well as digital signals—sine waves will do as well as square waves. This isn't a consideration for the project that we're working on now, but it does mean that you'll be able to use the same technique for experimenting with the transmission of voice and music. The circuitry's similar as well.

3. The remote control system will use DTMF signals.

There are lots of ways to encode the transmitted data, but the most useful one is the dual-tone multifrequency (DTMF) system developed by AT&T. There are standard, dedicated parts around to provide single-chip answers to our circuit problems and easy methods available to check the accuracy of our transmitted data.

One big advantage of going this route is that the project can be adapted to control things over the phone line. It's usually possible to generate the DTMF tones from the telephone keypad, but detecting them is a different story altogether. If you were building a system in which you could dial a number and then, when the connection was made, control things on the receiving end by hitting keys on the keypad, you'd still have to build the DTMF decoder.

A lot of the other encoding methods (FSK, for example) have advantages over DTMF (more available codes, and so on), but they don't produce tones that can be sent reliably over a pair of standard telephone lines.

The standard DTMF set provides only 16 codes; however, as we'll see when we get to the design of the decoder, there are ways around that limitation so you can get almost as many as you want.

4. Both the transmitter and receiver will be battery powered.

I can't imagine why anyone would design a remote control system where the transmitter had to be plugged into the wall but, for some reason, most of the receivers aren't battery powered. I suppose this has to do with the fact that most applications have the receiver buried deep inside something else—a television or VCR, for example—that wasn't designed to run off battery power.

Because we're designing this system with no particular application in mind, it's a good idea to make it as versatile as possible and that means being able to be battery powered. If you look back and examine the other things we've done earlier in the book, you'll see that all of them were built with CMOS parts. I prefer this family because it's inherently noise-immune, can run off flea power, and in the last few years has been upgraded in speed, versatility, and power.

The introduction of the 74HC and 74HCT chips means that you can now get pin-for-pin CMOS replacements for just about all the popular TTL chips. Not only that, but you'll find that they have almost the same fanout and drive specifications of their TTL equivalents. Once upon a time, a strictly CMOS design meant that you would have to modify your expectations to meet the inherent limitations of the CMOS parts that were available.

Fortunately, those days are over; and as far as I'm concerned, so are the days of TTL-only designs.

Basing the design of the receiver around CMOS parts means that the power requirements are going to be minimal and perfectly suited to a battery-based power supply. You won't be able to drive a constantly illuminated LED display or some other power-hungry device, but that's got nothing to do with the logic family.

These four criteria are really all we need. Now that you're getting to be an expert in the art of systematic design, some other ones may occur to you and, if they do, you should feel free to put them on the list. Hold off, however, on actually implementing them until we've finished the basic design because the ones on the list are going to be needed for any remote control system.

Adding your own ideas to the project is always a good thing. The only real way to learn anything is to do it yourself, but it's a better idea to add to something that already works than to modify something before it's up and operating.

I'd be disappointed if you only built the stuff in this book and then called it quits. You'll learn 10 times as much if you come up with your own ideas and, using the formal design method we've talked about, design your own stuff. The feeling you get the first time you successfully design something without anybody looking over your shoulder is better than any other one—well, almost any other one.

Don't worry about screwing stuff up. The popular myth is that you learn by doing, but that's not strictly true. The real scoop is that you learn by doing and making your own mistakes. Real understanding comes from figuring out what you did wrong and correcting it. If you start a project and (even though it's impossible) make it through to the end without making a single mistake, you've shortchanged the value of the time you spent on the project. You can learn more from a screwup than from a success.

Block diagram

As you undoubtedly know by now, the next step in the design process is to draw the block diagram of the project. We've been through this a couple of times together so you should already know how to pick the things that have to be put in separate boxes.

The remote control, as we've already said, is basically a simple project even though the circuitry is somewhat more complex than the other things we've built. This state of affairs is reflected in the block diagram in Fig. 9-2. No matter how much more experience you have on the bench, no matter how many more projects you get involved in, chances are you'll never find a case where the block diagram is as simple and linear as this one.

If you compare this one to the others we did earlier in the book, you'll see exactly what I mean. Here, all the arrows go from left to right, and there's only one arrow coming out of each box. You can make the diagram more complex if you want. The keyboard circuit can be broken down into the selector, clock, and encoder (more about this as soon as we get to the hardware); but you should already know about this stuff because we went through a similar design a couple of chapters back.

As we go through the design of the hardware for each section of the project, you should add more detail to the block diagram if you're not sure of the circuit or feel that things would be clearer with a richer, overall view of what's going on. Remember that the drawings you do here are going to live forever in your notebook; and while things may seem clear when you're doing them now, six months or a year down the road you might have a hard time both understanding and remembering all the details of everything that went into the design.

9-2 Block diagram of the remote control project.

It's a good exercise in design to draw block diagrams for any of the boxes shown in Fig. 9-2 that have more than a few active parts. The main reason for doing block diagrams in the first place is to give you an overall view of how the complete circuit is going to work. While it's possible to have too little detail, it's just about impossible to have too much detail.

Block diagrams are drawn to make your life easier, but there's no sense in doing them if they're not going to help you work your way through the design. Expanding an individual block and adding some more detail is the right thing to do if you can't look at a box and immediately know what elements are in it. Besides, paper is cheaper than silicon.

Keyboard

Even though we're not going to use the same circuit we did earlier in the book, you should still be able to design one of these without too much brain damage. There are lots of ways to make a keyboard happen; but as someone or other said, a keyboard is a keyboard is a keyboard.

For keyboards that don't have lots of switches, the best approach to the design is to use the same sort of scanning keyboard arrangement we did before as shown in Fig. 9-3. Some things don't change.

The main difference between this circuit and the one we did earlier is that we've specified that this one has to have at least 16 keys. We could cut this back to 10 but we've also said that we'll be using DTMF encoding, which will give us 16 possible tone combinations; thus, it's a bit shortsighted to decrease the minimum number of switches.

Wasting available output codes is always a loss and, in any event, the chance to do things differently should be viewed as a challenge, not a chore.

Since our criteria called for a minimum of 16 keys, we'll need a keyswitch selector that has at least 16 outputs. Just as we used a 1-of-10 decoder in the earlier circuit, we'll be using a 1-of-16 decoder in this one.

There are several chips that meet the requirements of having a low power consumption, 16 outputs, and the control pins needed to be able to cascade them if

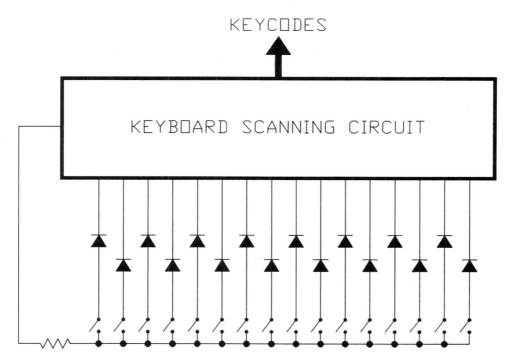

9-3 A scanning common-leg keyboard arrangement.

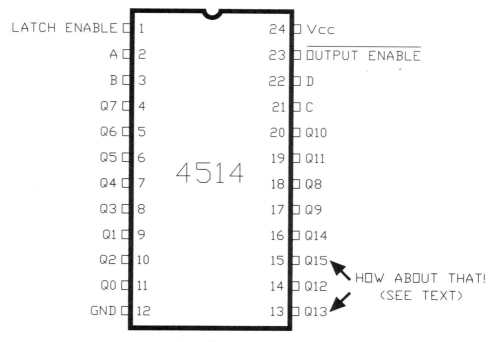

9-4 Pinouts of the 4514.

necessary. One such chip is the 4514, a 1-of-16 selector whose pinouts are shown in Fig. 9-4. It has a built-in latch; and although the outputs can't be tristated, there is a control pin to deselect all the outputs.

Outputs   The outputs are on pins 4 through 11 and pins 13 through 20. As we've seen in more chips than I can remember, there's absolutely no useful relationship between the pin numbers and the order of the outputs. The outputs are mutually exclusive in the sense that only one of them can be active at a time. The active output goes high and all inactive outputs remain low.

Address inputs   The address inputs are weighted in binary form. Because the chip is a 1-of-16 selector, there are four address inputs located on pins 2 (A), 3 (B), 21 (C), and 22 (D). When the IC is enabled, the binary data put on these inputs will determine which of the outputs will be selected.

Output enable control   The output enable control is on pin 23. When this input is made low, the chip behaves normally and the selected output goes high. If this input is made high, all the outputs are forced low regardless of the state of the rest of the chip's pins.

Latch enable control   The latch enable control on pin 1 is the key to the door of the IC's internal latch. When this input is made high, the output addressed by the word at the address inputs will be selected and the outputs will immediately reflect any changes in the inputs. If this pin is low, the latch will be closed and the chip will ignore changes and data at the address pins. The selected output will be the last one addressed while the latch enable control was high.

Other pins   The other pins are the power input on pin 24 and ground on pin 12. Since this is a CMOS part, it will operate happily at any voltage from 5 to 12 V. You can even go a bit beyond those boundaries but, if you do that, you'll be running the chip at the ragged edge and that's not a good thing to do.

I've noticed that an experienced designer can get confused whenever he reads a description of IC pins. This is understandable because just about every written description of an IC makes the device seem to be a lot more complex than it really is. While it's very true that nobody should design around an IC without having a databook open, you shouldn't ever be put off by the language used in describing how the chip works.

It's hard to believe, but semiconductor manufacturers design their products to be as useful and easy to use as possible. I'll agree that this isn't always immediately evident from the words in a databook, but that's more a function of how good the author is (there is a real, live human being somewhere who writes that stuff) and not always an indication of what the chip is like.

The 4514 and its companion chip the 4515 (exactly the same but it has outputs

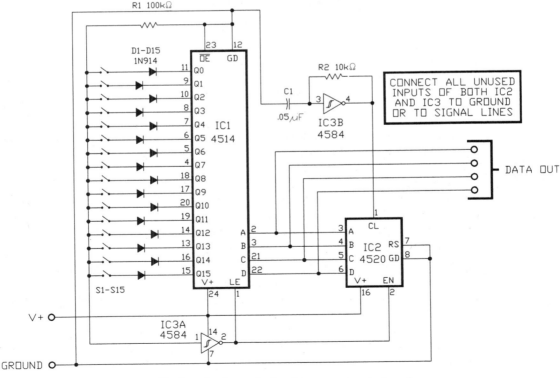

9-5   Keyboard schematic for the remote control system.

that are normally high and the selected output goes low) are easy to use. They're both good choices for switch selectors in a keyboard circuit because they have lots of outputs and a handy selection of control pins.

The basic keyboard circuit for our remote control system is shown in Fig. 9-5. Even though some of the components are different from the ones we used earlier in the book, the basic approach to the design is the same and, if you squint your eyes, you should be able to see the similarity. We're still using diodes to isolate each of the 4514 outputs to avoid having both a high and low on the "any key pressed" line at the same time.

We used a 4518 as the keycode encoder the last time around, but this time we're using a 4520. This is the hex counterpart to the BCD chip from the previous circuit, and that's absolutely the only difference between them. The 4518 counts in BCD and resets to a zero count after a binary nine (1001b) count is reached. The 4520 goes past this and continues on for the full 4-bit binary count of 16 (1111b or Fh) before it wraps back to zero.

The keyboard clock in this circuit is different from the 555 we had before. There's nothing wrong with the 555; and if you happen to have a bunch of them around, that's a good reason to substitute it for the clock shown in the schematic. The reason I'm using the 4584 is simply that I had some around and wanted to use them. Besides, it's a new clock circuit and a good indication of just how simply you can build a clock generator.

If you're using solderless breadboards, the placement diagram for this initial circuit is shown in Fig. 9-6.

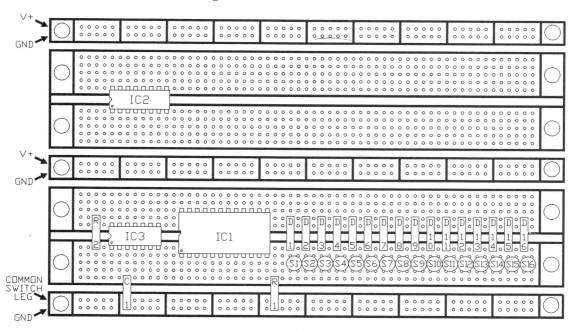

9-6 Placement diagram for the keyboard circuit.

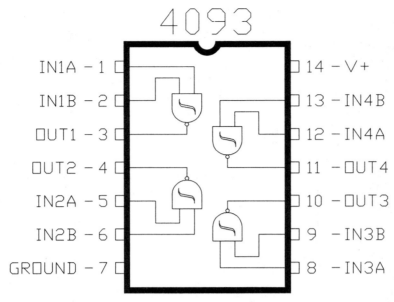

9-7 Pinouts of the 4093.

This clock, by the way, is a good one to keep in mind and is well worth putting on a separate page in your notebook. You can use any of the Schmitt trigger parts—I'm using the inverter, but you can also use the 4093 quad NAND gate Schmitt trigger shown in Fig. 9-7—and they'll work just as well.

The only hassle with these clock circuits is that the frequency is somewhat voltage dependent; but when you need a particular frequency, there are formulas you can use to get you in the ball park. Any fine trimming can be done by measuring the output and adjusting the value of the resistor. The frequency guideline formulas you can use for starters are as follows:

$$\text{(With a Vcc of 5 V)} \quad F = 1/0.5RC$$

$$\text{(With a Vcc of 10 V)} \quad F = 1/0.8RC$$

If you're using this clock circuit and expect the frequency to be relatively stable, remember that you also have to regulate the supply voltage.

When your application needs a really precise frequency, you're much better off with a 555 clock or, if you have to have crystal accuracy, something that's much more drift free and stable than either one of these.

The operation of this circuit is similar to that of the one we designed a bit earlier in the book, but there's one difference you should be aware of. If you examine the schematic carefully, you'll probably spot it. We went through a bit of brain damage last time around to make sure that the count in all the chips was maintained in sync.

To save you a bunch of page shuffling, the schematic I'm referring to is shown in Fig. 9-8. You'll notice that several resistor and capacitor pairs were used to reset all the chips at power-up. A good deal of attention was also given to have both chips, the 4017 and the 4518, reset to zero at the same time because the reliability of the circuit was dependent on having each output of the 4017 go active at the same time the correct corresponding number appeared on the output pins of the 4518.

When you compare this circuit to the one we're using now, you'll see that all the reset components and connections are missing. And if that wasn't enough, you'll find that this circuit—the one shown in Fig. 9-5—is more stable and reliable than the earlier one in Fig. 9-8 even though it's apparently missing all the things we added to the earlier circuit to increase reliability! Quite a mystery. But not really.

The reason for this apparent inconsistency is based on the fact that the circuits, while similar in function, are really different in design. The earlier circuit used a 4017 as the keyswitch selector, and that's where the difference comes in. We were using it to stop the count in the 4518, and we had to make sure that the switch we closed at the keyboard corresponded to the correct count on the output of the 4518. The 4017 outputs sequence from one to the next at the same rate as the 4518 increments its count, but there's no direct connection at all between the two chips.

Even though they're both being advanced by the same clock, there's no guarantee that the count in both chips will be the same. To see this, imagine that there's a glitch in the clock line, or in one of the chips, that causes the 4017 to

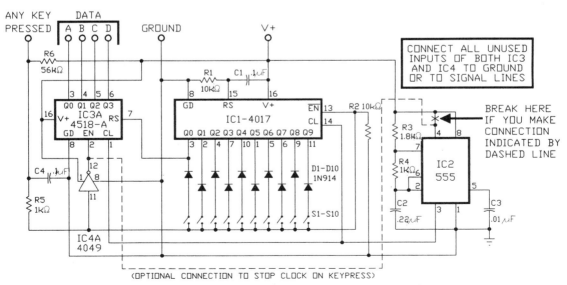

9-8 Keyboard scanning circuit from the keyboard project.

advance one count while the 4518 stalls for one count. The first thing to ask yourself is if this is even conceivable. The answer is yes.

We set up the earlier circuit so that sync is restored whenever the 4017 puts a high on its first output at pin 3. When this happens, a pulse is generated through the connection to the reset pin of the 4518, causing it to reset to zero at the same time. But this is the only time, assuming no keypress occurs, that the two chips communicate with each other. During the rest of the count—from one to nine—we're assuming that since both chips are being driven by the same clock at the same rate, they'll remain in sync until the count reaches zero again and the reset takes place again.

So, once again, ask yourself what happens if a glitch or some other piece of nasty business take place in the circuit. Is it possible for the two chips to get out of sync during that counting cycle? You bet your new pair of sneakers it is.

The sync gets restored at the beginning of each counting cycle, but who's to say that there isn't an occasional screwup during some of the cycles? You can be reasonably certain that everything's going to work correctly, but you just can't say that there's an absolute, ironclad, double-your-money-back type of guarantee.

Given all this, let's now take a look at the keyboard circuit we're using in the remote control. The difference in the design means that the uncertainty of the earlier circuit is no longer anything for us to worry about. A similar kind of screwup isn't possible—not because the sync between the 4514 and the 4520 is maintained more carefully, but because the two chips are directly communicating with each other during the entire counting cycle.

In the earlier circuit, both chips were being driven in parallel by the same clock. In this circuit, only one chip, the 4520, is being driven by the clock. The 4514 is being driven by the 4520, and it's the data being output by the 4520 that decides which 4514 output will be activated. The only 4514 output that can be active at any time is the one being addressed by the 4520—and that's why the whole issue of having to worry about maintaining sync is eliminated.

Before we go any further, there's one more difference between the two circuits that you should pick up on. In the earlier one, we had to be sure the mechanical switches were debounced. A noisy switch could cause multiple pulses on the "any key pressed" line. Because sync between the two chips was a concern, it's conceivable that a switch bounce could cause the kind of circuit glitch we were talking about a minute ago, and there might be a difference between the output state of the 4518 and the active output of the 4017 when the circuit stopped counting.

In the keyboard circuit we're using here, nothing whatsoever has been done to debounce the switches. Debouncing is simply not necessary—the circuit is inherently bounce-free in the same way that the two chips are always in sync. The best way to see this is to go through the operation of the circuit.

As long as no key is pressed, the 4520 counts repeatedly to 1111b and its outputs cause each of the 4514 outputs to go high in turn, one after another in sequence. When a switch is closed on the keyboard, nothing happens until that output is addressed by the 4520. As soon as the closed 4514 output is selected, the high causes the 4520 to freeze, the address of that switch (the data appearing on the outputs of the 4520) remains on the bus, and a keypress signal is put on the "any key pressed" line.

Now let's say that we're using some cheapie Neptunian switches and they bounce whenever they're pressed. No matter how lousy the switch is, no matter how badly it bounces, sooner or later the bouncing will stop and the switch will be constantly closed. After all, that's what switches do. Some do it better than others, but what makes one switch better than another is that it's quieter and more reliable. The state of all switches, those from Neptune and those from NASA, will be exactly the same a certain period of time after you press them.

If the switch bounces and, because of the bounce, happens to be in an open condition when its output is addressed, the count will simply continue and the circuit will test the state of the switch again the next time that particular 4514 output is selected. The bottom line is that if a switch is closed, the only data that can be frozen on the data bus is the address of that switch. It may take a few cycles of the counter before the switch closure is finally stable enough to be recognized, but only the address of that switch will appear on the bus.

Modulator

Now that we're generating valid keycodes, we have to take the data and convert it to whatever format we decided to use when we drew up the list of design criteria. In case you're suffering from short-term memory problems, you'll remember that we're going to use DTMF tones, and we're doing that for a couple of reasons.

DTMF, to start off with, is an accepted, well-established standard, and that means we can use this circuit in a variety of other applications unrelated to our present goal of designing and building a remote control system. There's no point going into this any further because we all know that DTMF tones are the basis of the telephone switching system, and just about the only places you can't reach by telephone are Metaluna (the lines were cut by the Zagons), Neptune (AT&T won't accept payment in herns, the Neptunian currency), and my house in the mountains (I won't give out the number).

When the DTMF standard was first introduced, every designer with earlobes got into the business of building circuits to generate the tones. Back then, digital stuff wasn't as diversified, so the whole job had to be done with analog oscillators. Everything would work but there were always problems with circuit size, frequency drifting, and other horrors. But as we all know, wine, children, and technology all mature with age.

The way things are today, only an electronic historian would sit down to design an analog-based DTMF generator. The friendly folks in the digital chip business have not only gotten into the DTMF generator business, but they've taken it over completely. You could design some digital circuitry to generate the tones, but this is one case where the semiconductor industry has made the job rather pointless. There are now so many different single-chip DTMF generators that are both cheap and available that designing your own multichip circuit to do the same job makes about as much sense as fixing lightbulbs.

I'm not saying you couldn't do it—I know you can do anything you set your mind on—it's just that, since they only cost a buck or so, there's not much point to it.

There's always a good reason for designing things from the ground up, but there are times when it pays to use a somewhat specialized IC to do a somewhat specialized job. There are so many IC manufacturers making such a wide variety of DTMF generators that they're running out of features to distinguish each one of them. Building your own DTMF generator circuit—even a digital one—means designing lots and lots of super stable oscillators, a complex input section to enable certain oscillators when certain data is presented to the circuit, and various output subcircuits to buffer and combine the generated frequencies that form the DTMF pairs.

Not a good way to spend March. And, what's even worse, you won't learn much doing it either.

When it comes to generating DTMF tones, there are so many dedicated parts around that you're much better off working out the requirements of your circuit and then seeing if there are any chips around that can fill the bill. In our case, the only special requirement we have is that the DTMF generator has to be able to be driven by the data from the keyboard circuit we've designed.

To be more specific, our keyboard circuit is putting out straight 4-bit binary for a total of 16 different possible numbers. As it happens, there are also 16 possible DTMF tone pairs that are made out of eight discrete tones which are listed in Fig. 9-9. The tones are divided into two groups called, appropriately enough, the high group and the low group.

The basic idea behind tone dialing was that a way had to be found to send signaling tones which wouldn't occur naturally as a result of speech, music, or some

DTMF TONES IN HERTZ	
LOW GROUP	HIGH GROUP
697	1209
770	1336
852	1477
941	1633

9-9   The DTMF frequencies.

DTMF TONES AND THE KEYPAD BUTTONS		
KEYPAD BUTTON	LOW GROUP	HIGH GROUP
1	697	1209
2	697	1336
3	697	1477
4	770	1209
5	770	1336
6	770	1477
7	852	1209
8	852	1336
9	852	1477
0	941	1336
#	941	1477
*	941	1209

9-10 The DTMF tone pairs and related telephone keypad symbols.

other everyday audio source. This was quite a problem, and it was further complicated by the fact that whatever the answer was, it also had to be something that would fit in the narrow bandwidth of standard phone lines—a restrictive channel only 2 700 Hz wide from 300 to 3 000 Hz.

The solution arrived at by AT&T was to use two tones for each signal. These tone pairs would be made from combinations of eight very precise frequencies (as shown in Fig. 9-10), and both tones would have to be present before the signal could be recognized as a valid DTMF tone. (Remember that the DT in DTMF stands for dual tone.)

Most standard telephone keypads have their switches arranged in a matrix as shown in Fig. 9-11. It can be a 3 × 4 matrix as in the case of a standard telephone, or a 4 × 4 matrix as in the case of a special telephone or signaler that sends data over the phone lines. In the early days of DTMF, the four extra keys in a 4 × 4 matrix were used by the military for . . . I don't know—military stuff.

If you look through databooks that list DTMF generators, you'll see that most of them are designed to be driven by standard, matrixed keypads. There are separate row and column inputs on the chip, and a tone pair is generated whenever simultaneous signals appear on a row and column input. Some of the chips have three column inputs for use with 3 × 4 keypads, and others have four so that a 4 × 4 keypad can be used to generate the full range of DTMF pairs.

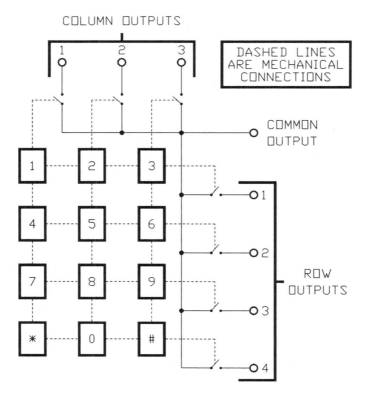

9-11 A matrixed telephone keypad.

We need a DTMF generator chip that can accept binary data, and there are several of those around. One of the easiest to use (always a good feature) is the TP5088 made by National Semiconductor (pinouts are shown in Fig. 9-12). This chip is designed specifically to be driven by 4-bit binary data so it can generate all 16 of the DTMF tone pairs. It also has all the features found in most of the other DTMF generators although, as we'll see, most of them are designed for use in building telephones, not remote control systems.

Because there are so many DTMF chips around, you may find it hard to get the particular one you want to use. It's always possible to buy 50 pieces, but getting stuff in onesies is always a problem and there's not much you can do about it. Several of the suppliers who carry the 5088 have minimum orders of $25 or $50 so you may be able to get the chip by including it in an order for other parts you need anyway.

If this isn't possible, there are other DTMF generators that can be driven by binary (the S2579 from AMI, for instance), and they'll be suitable for the project as well. If all this fails and the only chip you can lay your hands on is one that wants to see a standard keypad, you'll have to use that. It means dumping our keyboard circuit but, as they say, what can you do.

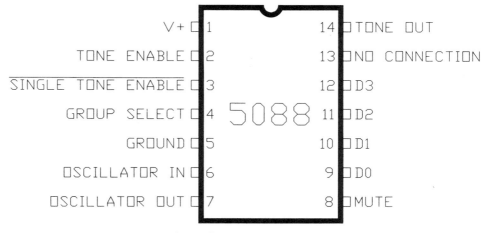

9-12 Pinouts of the 5088.

But don't give up hope too easily. The 5088 is a popular part, and with a bit of luck you'll be able to get one. When you do, this is what you should know about the inputs and outputs.

Data inputs   The data inputs are D0 on pin 9, D1 on pin 10, D2 on pin 11, and D3 on pin 12. These are standard binary inputs and are designed to be driven by standard bus strength signals.

Tone enable input   The tone enable input on pin 2 does two things for you. When it's high, the chip will generate DTMF tones, and when it's low the chip will not only be silent but will also be put in a low-power mode so that it only draws about 60 μA. This is great for a battery-powered circuit like ours.

Single tone enable input   The single tone enable input on pin 3 is mainly used for testing either the circuit you're building or the 5088 itself. When it's high, the chip will operate normally and generate DTMF tone pairs. If you make this pin low, the chip will only generate single tones from the high group or the low group.

Group select input   The group select input on pin 4 is only functional when the 5088 is set (on pin 3) to generate single tones instead of normal tone pairs. If it's high, the chip will generate high group tones, and if it's low . . . well, I'll leave it to your imagination.

Oscillator connections   The oscillator connections are the input on pin 6 and the output on pin 7. The 5088 has an onboard oscillator circuit designed to be driven by a 3.58-MHz crystal connected to these pins. The internal clock provides the base frequency which is divided down to generate the 16 DTMF frequencies. When the tone enable input is pulled low, the oscillator is disabled. This is one reason why the chip is able to go into a low-power standby mode.

Mute output   The mute output on pin 8 is a signal usually used to control the routing of audio to the handset when the chip is used as the DTMF tone generator in a telephone. When the tone enable input is low and no tones are being gener-

THE TRUTH TABLE OF THE 5088							
KEYBOARD BUTTON	DATA INPUT				TONE ENABLE	TONES OUT		MUTE OUT
	D3	D2	D1	D0		HIGH	LOW	
X	X	X	X	X	0	GND	GND	GND
1	0	0	0	1	_/‾	697	1209	O/C
2	0	0	1	0	_/‾	697	1336	O/C
3	0	0	1	1	_/‾	697	1477	O/C
4	0	1	0	0	_/‾	770	1209	O/C
5	0	1	0	1	_/‾	770	1336	O/C
6	0	1	1	0	_/‾	770	1477	O/C
7	0	1	1	1	_/‾	852	1209	O/C
8	1	0	0	0	_/‾	852	1336	O/C
9	1	0	0	1	_/‾	852	1477	O/C
0	1	0	1	0	_/‾	941	1336	O/C
*	1	0	1	1	_/‾	941	1209	O/C
#	1	1	0	0	_/‾	941	1477	O/C
A	1	1	0	1	_/‾	697	1633	O/C
B	1	1	1	0	_/‾	770	1633	O/C
C	1	1	1	1	_/‾	852	1633	O/C
D	0	0	0	0	_/‾	941	1633	O/C

9-13   Truth table of the 5088.

ated, the mute output will be low. When the tone enable input is high to allow tone generation by the chip, the mute output will be an open collector (turned off). In a normal telephone application, the signal from the mute output will be used to prevent you from hearing the tones when they're being generated.

Tone output   The tone output on pin 14 is no surprise here—where the tones will appear. There's an internal transistor on this output; and when no tones are being generated, the transistor is turned off to cut down on the chip's power consumption.

Other pins   The other pins are power on pin 1 and ground on pin 5. The only thing to notice here is that these are not the standard locations for these pins. If you overlook this fact and put power and ground on another pair of pins, the only sound you'll get from the chip will be a momentary death rattle.

You can see how the state of the control pins affects the operation of the 5088 by studying the table in Fig. 9-13. We won't be using some of the chip's functions, but you should have this information if you're going to be using the chip in this circuit or in any other circuit you build.

I don't have to remind you about databooks, do I?

Whenever you build a circuit described in a book, you'll often find that the author tells you that it's really easy to do. Well, as we all know, sometimes that's true and sometimes it's not true. Calling something easy is as much an indication of a state of mind as it is a measure of the difficulty of a project.

Connecting the 5088 to our keyboard is easy. Really. As a matter of fact, if you can think of any other circuit addition you've done that's easier, I personally will come to your house and wire the 5088 into the breadboard. Lab tests conducted in college departments (funded by the government at outrageous cost) have shown that gorillas born in the wild can wire 5088s up in 14 minutes flat. Domestically born gorillas needed half an hour, but they're in the union.

The schematic for this circuit addition is given in Fig. 9-14.

Put the 5088 on the board as shown in Fig. 9-15 and connect the crystal across pins 6 and 7. The only other connections you have to make are to wire the chip's data inputs to the bus that's coming from the keyboard. It's also a good idea to make the connections from the single tone enable and group select pins someplace where you can get your hands on them so you can easily put the chip into its single-tone mode. If you have any trouble, you'll want to be able to have the chip generate individual frequencies.

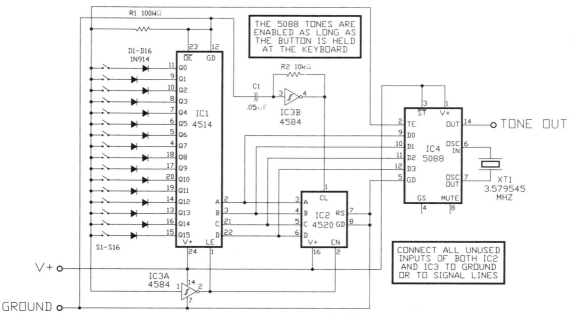

9-14 Schematic of the basic DTMF generator.

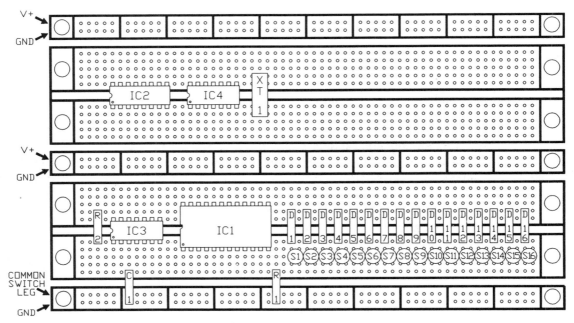

9-15   Placement diagram for the DTMF generator.

Transmitter

This is the last part of the circuit we have left to design, and you'll remember that we specified infrared as the way we were going to get the data from the transmitter to the receiver. This isn't a difficult subcircuit to design, but the particular configuration will depend on the DTMF generator you use. There's a variation in the strength of the signal you can get from chip to chip.

It's a bit much to ask any DTMF generator to provide enough output current to directly produce infrared; thus, the transmitter you use should have the capability of amplifying the signal. How much amplification you'll need is also chip dependent so we should design the transmitter in such a way that the amount of gain it provides can be set to match the level coming from the DTMF generator.

The most straightforward approach to the design is to use an op amp because this is a quick-and-dirty way to boost small signals to a more usable level. Of all the integrated circuits on the market, probably the largest family is the collection of op amps made by most of the manufacturers in the industry. We're not asking a lot from an op amp so the choice of which part to use is pretty much open.

As you can see in the schematic of Fig. 9-16, I've put an LM386 on the board. It's a virtually bulletproof IC that can put out about 300 mW into an 8-Ω load. It's really designed to drive a small speaker but will work perfectly here as well. The overall gain of the chip is controlled by the resistance between pin 1 and pin 8. The 386 has an internal 1 350-Ω resistor across these pins to give the chip a gain of

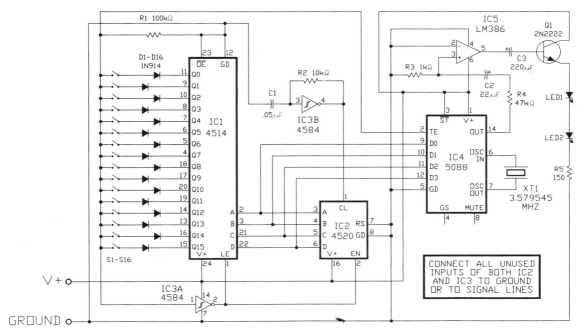

9-16 Transmitter schematic.

about 20 dB. The maximum gain you can get from this chip is about 45 dB by bypassing the internal resistor with an external 10-μF capacitor. Unless you're using a DTMF generator chip I never heard of, the basic 20-dB maximum gain should be more than enough.

By wiring up the circuit as shown in the schematic, you can trim the gain by playing with the value of the resistor in series with the output of the 5088 (or whatever DTMF generator you're using). You can even eliminate the resistor completely since the 5088 is capacitively coupled to the 386. Just remember that too strong a signal at the input of the 386 might result in an overload that will introduce some distortion to the final signal. Better a bit too little than a bit too much.

The output of the 386 has a substantial dc component; thus, the signal is routed through a capacitor, as shown in Fig. 9-16, before going out to the base of the 2N2222. The transistor is used as a simple switch to control the LEDs connected between its emitter and ground. The output of the 386, therefore, will produce a modulated light output from the LEDs. Infrared LEDs will generate infrared and plain LEDs will generate visible light.

There's nothing particularly tricky about this circuit; and if you add the components to the board as shown in Fig. 9-17, just about the only thing to watch out for is that you get the orientation of the polarized components (the capacitors and LEDs) correct. You won't have this problem with the capacitors because they're usually clearly marked, but LEDs are always a different story.

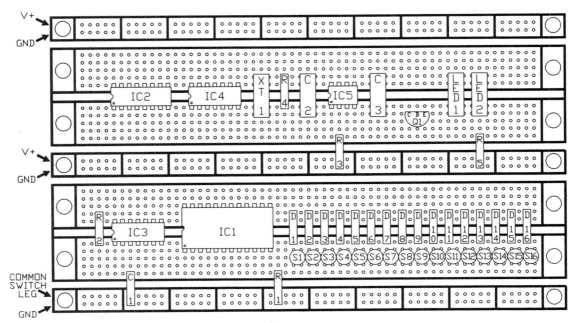

9-17 Placement diagram for the transmitter.

The best way to make sure the LEDs are facing the right direction when you put them on the board is to use two regular LEDs at first. When you get them to light, replace one of them with an infrared LED and, if the visible one still lights, you know that the orientation is correct.

Since we want as much infrared power as possible, you should use two LEDs, and both of them should be of the high-efficiency type. You may not believe this, but these LEDs—the exact ones you want for the project—are available in blister packs from Radio Shack. The same is true of the LM386. When you find them there, buy a few extra ones because if experience is any teacher, these parts won't be there when you go to get replacements for the ones you blow up. That's the way it is.

Once you've got the circuit completely wired, you should see the LEDs light every time you close a switch on the keyboard. Don't put the infrared LEDs on the board in place of the regular ones until you have no doubts that the circuit is working. Optimism is a necessary ingredient for any designer, but there's a big difference between being optimistic and being foolish.

Let me tell you, there are more than enough unexpected sources of potential aggravation. The goal is to keep them to a minimum, not go out of your way to find them.

Oops!

If you tried and verified the operation of each of the subcircuits we built, there's just no way you can be having a serious problem with the final circuit. The transmitter should be working, and the only way you might have a problem is if you pulled something loose or made a bad connection by putting one of the wires in the wrong place. The advantage of solderless breadboards is that they make it easy to add connections to a circuit. The disadvantage is that all the holes look the same—especially when you're tired.

One neat way to verify the operation of the circuit is to feed your stereo's line input or a set of headphones with the signal at the base of the 2N2222 output transmitter. Watch out for the volume setting and keep the earphones away from your ears because the level of the DTMF tones from the circuit can be very loud indeed. If you do this, don't take the signal directly from the output of the 386 at pin 5. There is dc there that may damage your stereo, and loading the chip at that point will damage the IC.

This circuit is a really easy one to troubleshoot with the famous wham-bam method. Just start with the output of the 5088 and work your way forward. If there's nothing there to begin with, use your logic probe or multimeter to check that the keyboard is scanning, that you have data on the bus, and that the circuit freezes data on the bus when you press a key.

If the keyboard is working, compare the connections you've made to the 5088 pins against the schematic. Remember that it's possible to disable the 5088 by a bad connection to the tone enable input; and since we're controlling that with the "any key pressed" line, a bad connection here will keep any tones from being generated.

The next part of the project is to build the receiver, but there's no point to starting that until and unless the transmitter is working correctly. I haven't seen your breadboard, but mine is sitting on the bench exactly as shown in Fig. 9-17, and it's doing what it's meant to do. There's no reason why yours should be any different—unless you did something differently.

A slick trick

Both CDs and, to some extent, the newer cassette tapes have a tremendous dynamic range. Most newer amplifiers have limiters built in to keep excessively loud signals from clipping or even damaging the output stages of the amplifier or the speakers themselves. The best way you can automatically limit an amplifier is to turn it down even before the amplifier can process the signal and go into overload.

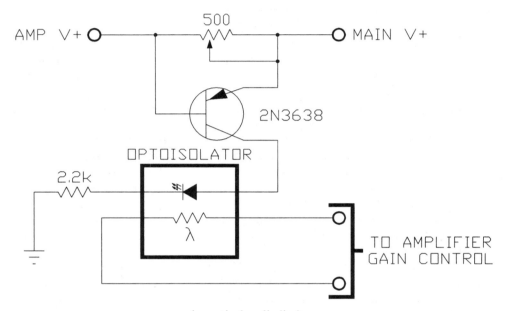

An optical audio limiter.

The resistor sits on the amplifier's power line, and a voltage develops across it that is directly proportional to the amount of amplification it has to do. The current flow resulting in the transistor drives an LED/photoresistor combination to cut the amplifier gain so there's no way it even has a chance to clip the signal.

You can use the potentiometer to change the sensor resistance and vary the level at which the limiter will start trimming the amplifier gain.

10
Remote control #2
Making it happen there

I don't know why it's true, but everyone I've ever met who has gotten involved in electronics is also a science fiction freak. Not everyone reads it, but everyone goes bananas over the movies. Understand what I'm talking about here. Science fiction movies aren't the ones that involve lots of blood and gore, or slimy creatures that seem to be made with their internal organs on the outside. (These are the creatures who are always eating the male inhabitants of the same town on the California coast and also overcoming astounding anatomical difficulties by violating the women—all of whom are gorgeous—though I don't understand, from a biological point of view, why they'd bother doing it or, from a strictly mechanical point of view, how it would be possible in the first place.)

Science fiction is the theater of the possible future, and the only movies that fit that definition are the "let's go to Mars for lunch" type. The only way I can accept the idea of human beings as alien lunchmeat is if the things doing the eating had to travel here from another galaxy—or at least another planet.

If you've never seen the old *Flash Gordon* series made in the late thirties with Buster Crabbe in the title role, you're really missing something. Three of them were made, and each was something like 15 cliff-hanging chapters long because they were originally shown in the movies at Saturday matinees. They were also shown on television (that's where I saw them), and I know for a fact that they were condensed into feature-length movies. Even though the compilations are missing a lot of the really good stuff (like all the cliff-hangers), they're available on videotape and you should get your hands on them.

The three serials and the compilation titles are:

- *Flash Gordon* – "Spaceship to the Unknown"
- *Flash Gordon's Trip to Mars* – "Deadly Ray from Mars"
- *Flash Gordon Conquers the Universe* – "Purple Death from Outer Space"

While they're a long way from being the most inspired handling of the original series' footage, they're still better than nothing. At least you'll be able to get a taste of what the whole thing was like.

The best actor in the series (all three of them) was undoubtedly Charles Middleton, who had the part of Ming the Merciless. As far as I'm concerned, he was a classic bad guy—smart, neat looking, and just unbelievably twisted.

You might well wonder why I'm talking about this stuff because it's a bit removed from the subject at hand—building the receiver portion of our remote control system. That's a good question and deserves an equally good answer.

Everybody makes associations that become permanently embedded in memory. From the first moment you make the mental connection, the two things are inextricably joined together. You'll never be able to think about one of them without thinking about the other as well.

Whenever I run across the two words *remote control*, I always think about one line from one scene in the *Flash Gordon* serials. It's in the last episode of the second serial. Flash Gordon, Dale Arden, and Doctor Zarkhov are returning to Earth in Zarkhov's rocket ship and are talking to Doctor Gordon (Flash's father) on Earth. Zarkhov has a problem with his rocket ship and tells Doctor Gordon (a big-shot scientist) that there's a load of electrical interference coming from Earth and, unless it's stopped, he won't be able to safely land his ship.

Doctor Gordon, without missing a single beat, turns around to one of his assistants and says, "Turn off all the electricity on Earth." His assistant rushes over to a wall panel covered with all sorts of scientific-looking dials and throws a huge switch. The scene cuts back to the rocket ship. Zarkhov turns to Flash and says, "Well, that takes care of that problem." Talk about remote control!

The remote control system we're building probably won't ever be able to match that, but if you can find where that switch is located, I'll be happy to design the system to take control of it. Just drop me a line when you've found it.

Remote control receiver

If you think about it logically, the design of the receiver should be exactly the opposite of the transmitter design we finished at the end of the last chapter. Instead of generating DTMF tones and sending them out as pulsating beams of infrared light, we want to take the generated infrared and convert it back to DTMF tones. Logical analysis tells you that doing one operation is just the opposite of doing the other one. Well, unfortunately, that's not quite true.

The block diagram for the transmitter and receiver we want to build is shown in Fig. 10-1. There's symmetry there because, on the most conceptual level, the two sections are doing similar, though opposite, jobs. When we get into the details of the receiver sections shown in the block diagram, however, you'll see that it's a lot simpler to generate DTMF tones than it is to regenerate them. But we're getting ahead of ourselves.

10-1 Block diagram of the remote control system.

The additional design problems of reconstituting the transmitted DTMF tones will become apparent later when we get to the design of the demodulator. You'll see that it's a lot easier to combine tones than it is to separate them.

But let's start at the beginning.

Sensor

The first section we need to build in the receiver, no matter what we do later in the project, is something that can convert the infrared light from the transmitter into board-level electronic signals. Once that's done, we can get on with the design of the remaining subcircuits that form the completed receiver.

Detecting infrared is no big deal because there are phototransistors that have good sensitivity in the infrared band. The problem with all of them is that they're also sensitive to ordinary visible light. If we expect them to respond only to our transmitted infrared signals, we're going to need some method of making sure that the only kind of light they see is infrared light.

The most efficient way to do this is to use an optical filter that only passes infrared light. We can house the receiver in some sort of lighttight enclosure with a small filter-covered window and set the box up so the window gets pointed toward the general direction of our transmitter.

And so we come to the subject of infrared filters. Really efficient infrared filters are expensive. The first place to turn for optical filters is a Kodak handbook; and if you get one, you'll find that the Wratten No. 87 filter is exactly what you need. A 2-inch-square filter will set you back about 10 bucks. This may or may not be a lot of money but, just on general principles, the idea of shelling out 10 bucks for a small piece of thin plastic gel rubs me the wrong way.

There's a good alternative, however. If you can get your hands on a small piece of unexposed, developed Kodachrome film, you'll find that it makes a good substitute for the more expensive No. 87 filter. The best place to find a piece of film like this is in a box of Kodachrome slides. You'll generally get back the whole roll; and while the pictures are in cardboard mounts, you'll probably also find some blank pieces of film in the box as well. These pieces are the head and tail of the roll; and because they've never been exposed to light, they're blank. They're also good infrared filters.

If the labs you use to develop your Kodachrome don't return these pieces to you, ask them if they have some around. If they don't or won't, run off a few blank exposures on your next roll of film. Just keep the lens covered and shoot a couple of exposures. You may be wasting film, but you'll get the blanks back mounted conveniently in handy cardboard frames. Not bad.

Before we leave the subject of filters and go on to the design of the electronics, it's important for you to realize that the infrared sensor must be optically isolated from the open air with an infrared filter. Remember that the phototransistor is as sensitive to visible light as it is to infrared; and from the point of view of our remote control system, anything in the visible spectrum is noise. Leaving the phototransistor open to all light means that it can see the same stuff that you see and that means the phototransistor is going to be conducting, to some degree or another, all the time.

You have to make sure that the only time the phototransistor gets turned on is when it's hit with infrared light. The front-end design of the receiver is based on the idea that the signal-to-noise ratio (infrared light to visible light) is very high. As we get into the hardware, you'll see that it hinges on the assumption that the only signals coming from the sensor are due to the infrared light from the transmitter. Keep that in mind.

The schematic for the front end of the receiver doesn't involve a lot of parts. I'm talking now just about the phototransistor. As you can see in Fig. 10-2, it would be hard to find anything that's simpler. The phototransistor I'm using is a TIL414. There are other ones that are better, but this one is available from just about every supplier in the known universe—even Radio Shack!

The signal from the phototransistor is too low to be used directly so we have to amplify it. How much amplification is necessary will depend on the phototransistor you're using. For most phototransistors the schematic shown in Fig. 10-3 will provide enough gain to drive the rest of the electronics in the receiver. A single op

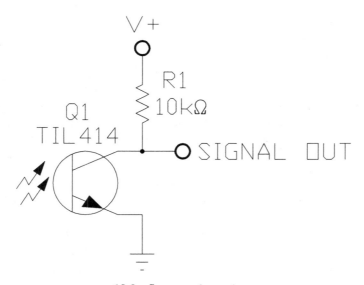

10-2   Sensor schematic.

10-3 Basic amplifier schematic for the sensor.

amp circuit should give you an adequate signal level, but this is something you'll have to determine on your own because everything depends on the parts you're using on the board.

I built the circuit on solderless breadboards, laid the parts out as shown in Fig. 10-4, and was able to get an acceptable level at the output of the op amp with the values shown in the schematic. The 741 is really designed to be used with a split voltage supply; but if you're not using it to amplify quality audio, you can operate it off a single-ended supply as shown in the drawing.

There may be some purists among you who feel that anything designed to work off a split supply should have a split supply. If you happen to be one of those people, you can use the negative supply described earlier in chapter 2 to generate the voltage. As a matter of fact, if you've never done it before, this is as good a chance as any for you to try it out.

Although you can use just about any op amp you want in this part of the receiver, the 741 is a good choice because it's cheap, available, and doesn't require any special handling to work reliably. In order to get an adequate level out of it,

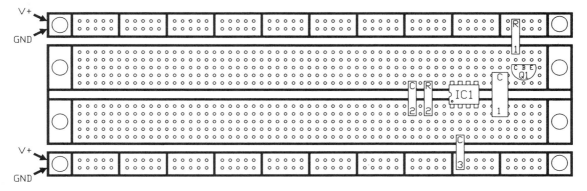

10-4 Placement diagram for the basic amplifier.

you'll notice that the gain (set by the ratio of the feedback resistor to the input load resistance) is extremely high—as a matter of fact, close to the open-loop gain of the op amp. This is the gain you would get with an infinite value for the feedback resistor—in more practical terms, no resistor there at all.

The easiest way to determine whether you're getting enough signal out of the amplifier is to connect the op amp output to your stereo's line input through a 10-μF capacitor. When you point the transmitter at the phototransistor and press one of the keys, you should hear the tones on your stereo. If you have an ac voltmeter, you should be able to get at least 1 Vrms from the op amp when the phototransistor is receiving a signal from the transmitter. If you're getting less than that value, you're going to need more amplification for the rest of the receiver circuitry to operate properly.

You can do this by changing op amps—a TL084 is a good choice—to try and get more level, but chances are you're better off adding a one-transistor preamplifier to the circuit as shown in Fig. 10-5. If you do this, you'll be dropping the gain in the op amp. Since the amount of gain you can get from this circuit is substantial, use the potentiometer in the feedback loop to set the output to an appropriate level.

If you don't have any way to measure the ac voltage at the output of the op amp, you can get by just listening to it on your stereo and making sure that there's enough level to provide a clear signal with the stereo's volume control set to what you consider a normal setting.

I know this is far from being scientific, but there's an awful lot of latitude in the circuitry for the rest of the receiver, and it's been designed to operate happily with an audio signal in the range of 0.75 to 1.5 Vrms. This is usually the level you'll get out of stuff like tape decks and CD players so if the stereo volume control is set the same for your audio junk as it is for the DTMF sensor, you should be in the ball park.

You have to have this part of the receiver working properly before you go on to the rest of the circuit. Take as much time as you find necessary to get both an adequate output level and a clean DTMF signal before you go on with the design. It's really tempting to say that something is good enough and move on, but that's a mistake that can cause real problems later on. It usually takes only a little bit of

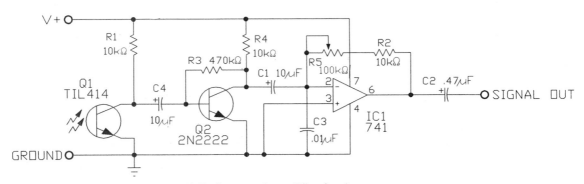

10-5 Improved amplifier for the sensor.

extra effort to make something work above the bare minimum needed for the circuit to function correctly, and that extra bit of margin is exactly what's meant by circuit reliability.

Remember that

Good enough is only good enough.

Circuits that produce signals that are only adequate are only adequate circuits.

The circuit in Fig. 10-3 (op amp only) will probably work for you, but you're running the op amp all the way at one end of its gain range. While there's nothing electronically wrong with that, I always like to keep something in reserve. Good circuit design means that all components are running in the middle of their rated maximums and potentiometers are always set in the midpoint. If you needed a 5-V supply and were going to draw a constant current of 1 A from it, you'd never consider designing something in which a 1-A draw was the maximum you could get from the supply. That's kind of like having a car that would only do 55 miles per hour and blow up if you tried to do 60. Bad design is bad design.

You're the one building the circuit so the choice is yours; but if I were you, I'd use the circuit shown in Fig. 10-5. That's the one I finally used. Even though I could have easily gotten by without the extra transistor stage, I felt more comfortable having it on the board. I laid out the components as shown in Fig. 10-6 and was able to turn the op amp potentiometer down to about its midpoint.

When you spend too much time designing digital circuits, it's easy to forget that some circuits can recognize values other than zero and one—nothing or everything. Not too many people know that when Klant Zorch designed the first version of the Neptunian warp drive, he fell into the same trap. The first Neptunian spaceships could either stand still or travel at 14 times the speed of light—nothing else. Once the problem was realized, a product recall was issued, but none of the owners had survived the acceleration. Live and learn.

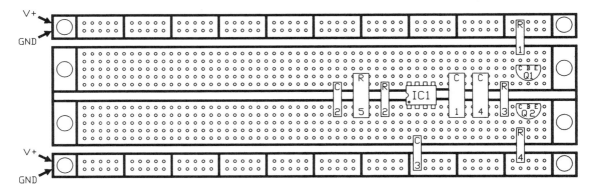

10-6 Placement diagram for the improved amplifier.

Demodulator

The demodulator is the heart of the receiver. Absolutely nothing is going to happen unless we can convert the received DTMF into codes that tell us which one of the 16 buttons was pressed on the transmitter. The problem in front of us is that we're faced with an audio signal from the op amp that has two frequencies in it—one from the low group and one from the high group. How well our circuit is going to work is a direct function of how well we can separate the two frequencies and be sure which ones have been sent.

You'll remember that we used a dedicated DTMF generator when we had to produce the tones in the transmitter. We did this because it's a pain in the neck to do it with discrete components, and the job would have taken a board full of silicon. By using something like the 5088, we cut the component count, circuit complexity, bench time, and brain damage.

The same considerations are a factor in the design of this part of the receiver. There's nothing stopping you from using a phase-locked loop like the LM567 to decode the DTMF frequencies. Once upon a time, all the DTMF decoder boxes in the galaxy were done like this. The basic circuit is fairly simple. As you can see in Fig. 10-7, it doesn't take a lot of components to put one together. The LM567, or any other phase-locked loop, can be set to detect a particular frequency, and its output will go high when the set frequency shows up at its input.

The down side of this design is that you have to duplicate it eight times because there are eight discrete DTMF frequencies. You're also using a couple of resistors and capacitors with each one of the eight subcircuits, and the values for each set have to be calculated to match each of the different DTMF tones. Lots of

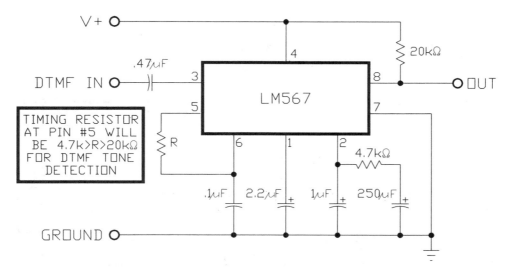

10-7   Detecting DTMF frequencies with a 567 phase-locked loop.

components means a lot of possible component drift and other hassles that will affect the overall reliability of the circuit.

But let's say that we're the kind of people who like beating our heads against the wall (figuratively only, I hope), and an excessive amount of circuit complexity is the kind of thing we love to sink our collective teeth into. So there we are—the math is done, the circuit is designed, and we're looking at eight outputs, two of which will be active when a DTMF tone pair is sent by the transmitter and amplified by the sensor.

Think about what's next. Knowing which frequencies were sent by the transmitter isn't the end of the job. In fact, we don't care about the particular frequencies. They're just a way of encoding the 16 buttons on the keypad. What we're after is a way to know which of the buttons is being pressed—we want the keycode that generated the tones, not the tones themselves.

This leaves us with an interesting (to some) problem in digital circuit design. We'll need a bunch of two-legged gates—16 of them, in fact—that will monitor the various high group/low group combinations and tell us which key was pressed on the transmitter. A circuit like this is certainly within reason, but it's far from being a reasonable circuit. Designing a DTMF decoder this way has to result in a circuit whose tone detection is less than reliable with a keycode decoder section that's going to have to be built from a lot of basic MSI gates.

If you put your mind to the problem, you might be able to work out a simpler circuit arrangement, but that's something you'll have to do on your own. Problems like that usually make my teeth hurt and give me migraines—especially when I know that they're not necessary. Dedicated ICs are a lot better to use.

All of the DTMF decoder chips want to see the high- and low-group tones on separate pins. This means we have to split them up before we can decode the tones and wind up with our goal—knowing which one of the 16 possible keys was pressed on the transmitter. Separating the high- and low-group tones is a straight problem in filter design, and the usual way to handle it is to use precise high- and low-pass filters to do the job. Commercial equipment uses ceramic filters which are somewhat costly and not too easy to find. We're going to use a chip that not only makes the job easier, but also has outputs perfectly tailored for the DTMF decoders.

My favorite group splitter is the S3525 from AMI. There are probably other chips around that do the same job, but this one's been on the market for a number of years and has a bunch of features which, in addition to its basic job, make it a good choice for our project. The chip may seem complex when you first look at it; but after we go through the pins, you'll see that it's really simple to use. As we go through the pin descriptions, keep referring to the block diagram of the chip (Fig. 10-8) because each word in the description is only worth a thousandth of the picture.

The pinouts are shown in Fig. 10-9.

Reference voltage   A reference voltage is available on pin 2. This voltage is generated by the resistive divider you can see in the chip's block diagram. The value of the voltage on this pin is equal to half the difference between the supply

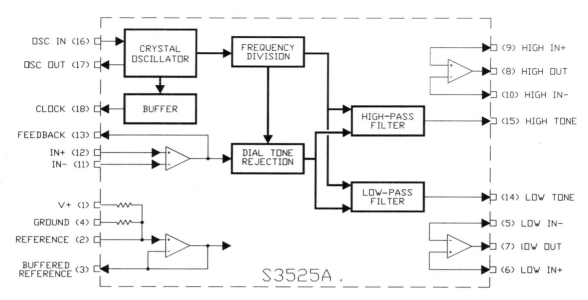

10-8 Block diagram of the 3525.

voltage and ground. Since you normally have the chip ground at system ground, the reference voltage is normally just half the supply voltage.

Buffered reference voltage The buffered reference voltage appears at pin 3. As you can see in the block diagram, this is the same value as the reference voltage on pin 2.

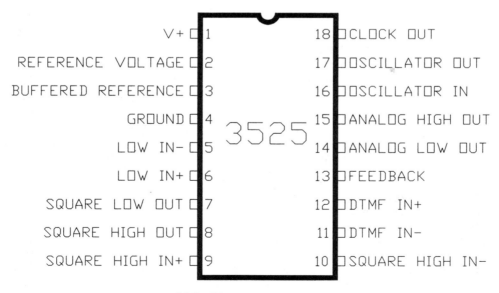

10-9 Pinouts of the 3525.

Uncommitted comparator inputs   The uncommitted comparator inputs for the IC's two internal comparators are on pins 5 and 6 (inverting and noninverting, respectively) for one comparator, and pins 10 and 9 (inverting and noninverting, respectively) for the other comparator. These two comparators are used for the final squaring up of the outputs of the high- and low-group filters in the chip. The easiest way to think of them is as the output stages for each of the groups after they've been separated by the chip's internal filters.

Uncommitted comparator outputs   The uncommitted comparator outputs are brought out to pin 7 (the first comparator) and pin 8 (the second comparator). After you have the chip properly configured, these pins will have square-wave versions of the high- and low-group frequencies.

Input amplifier controls   The input amplifier controls are on pin 11 (the inverting input), pin 12 (the noninverting input), and pin 13 (feedback). Having these pins available on the chip means you can tailor the input gain to match the signal source with the input level requirements of the 3525.

Filter outputs   The filter outputs are on pin 14 (the low group) and pin 15 (the high group). These are the analog outputs of the filters and can be routed directly to equipment that wants to see analog-type sine waves or can be fed back into the chip's two comparators to be squared up for equipment that wants digital-type square waves. This is one of the 3525's extra features.

Crystal inputs   The crystal inputs are on pins 16 and 17. The master clock for the chip can be set by connecting a 3.58-MHz crystal and 10-MΩ resistor in parallel across these pins. The internal dividers in the chip produce all the frequencies the chip needs to detect and separate the high- and low-group frequencies.

Crystal output   The crystal output is available on pin 18. This is a thoroughly buffered version of the chip's base frequency—the frequency stamped on the crystal. This means that any other chip in the circuit that needs a similar crystal frequency can get the clock pulses from this pin. Since most DTMF decoders also use a 3.58-MHz crystal for their basic clock, only one crystal will be needed to build the complete circuit.

Other pins   The other pins are pin 1 (power) and pin 4 (ground). This is a CMOS chip but—and this is important—it needs at least 9 V to operate. Note that both the power and ground pins are not at the standard locations. Should you forget this and put power and ground on pins 9 and 18, respectively (the standard locations), the chip will be history; and just about the only thing you'll be able to do will be to paint eyes on it and put it on the bottom of your fish tank.

The 3525 is handy because it makes really easy something that would ordinarily be a pain in the neck to design. What's most interesting about this chip, and what makes it such a pleasure to use, is how the AMI designers laid out the chip's internal circuitry. Most dedicated ICs are designed with the big brother "Take the Bus and Leave the Driving to Us" kind of attitude. Chips designed with this philosophy are usually picky about input levels and voltages; and if you want to use them in a slightly different fashion, you'll have to use a different IC. They do one job in only one way; and if you can't live with it, you'd better look around for something else.

The 3525 gives you access to most of the internal workings of the chip and lets you pick the input level as well as the kind of output you're going to get. Take another look at the chip's block diagram and notice that if you decide to use analog outputs, you've got two uncommitted op amps available for some other use in your circuit—and you can use them for anything you want.

Before we get into the actual design of the hardware, there's another chip we have to look at because the 3525 only does part of the job we need done. Remember that the 3525 is only ("only" is a very relative term—try doing the same job with discrete components and you'll see what I mean) separating the high- and low-group tones for us. This is still a bit away from getting back the original key-code generated in the transmitter.

Once we've separated the high- and low-group tones, we have to do the reverse job we did in the transmitter—convert them back to a 4-bit binary code. This could be a mammoth undertaking; but luckily for all of us, it's such a common thing to do in a circuit that there are dedicated ICs designed just for that purpose. Several semiconductor manufacturers have jumped into the market so there's a wide variety of available chips.

Every large manufacturer works toward finding its own particular part of the market, and Mostek seems to have made a specialty of DTMF decoder chips. Mostek has several of them in its catalog, and two of the chips are perfectly suited to our project. I'm using the 5103 but the 5102 is almost exactly the same chip—you can use either of them but if you have a choice, get the 5103. The only difference between the two chips is that the 5102 can accurately detect the DTMF tones on noisier lines than the 5103 (18-dB signal-to-noise tolerance on the 5102 as opposed to 14 dB on the 5103). The 5103 has a lower current consumption than the 5102—2 mA versus 5 mA—so there are advantages to both.

Because the two chips are so similar, especially from a functional point of view, we'll go through the pins of the 5103 and point out any differences between the two along the way. There aren't many differences and the pin locations are the same. The two ICs are pin-for-pin equivalents and, for our project, it makes no difference whatsoever which one you use. The pinouts are shown in Fig. 10-10.

Crystal inputs   The crystal inputs are on pins 2 and 3. The chip wants to see a 3.58-MHz crystal on these pins to use for its internal master clock, although it can take a 3.58-MHz clock input from the source on pin 2. If you do this, pin 3 can be left floating.

Strobe output   The strobe output on pin 4 goes high as soon as the chip decides that valid DTMF data has been put on its inputs. The 5102 is a little bit slower than the 5103 in detecting tones (it needs 33 ms as opposed to 30 ms for the 5103), but what's 3 ms between friends?

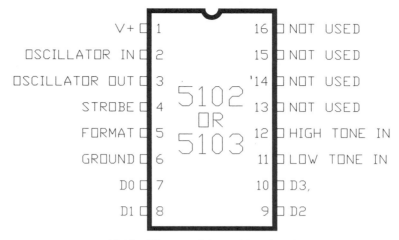

10-10 Pinouts of the 5102 and 5103.

Format input   The format input on pin 5 determines how the DTMF tones are going to be decoded and also serves as the chip's tristate control. If this pin is high, the outputs will give you 4-bit binary; if it's left floating, you'll get row and column codes; and if this pin is made low, the data outputs will be tristated.

Data outputs   The data outputs are on pins 7 (D0), 8 (D1), 9 (D2), and 10 (D3). The format of the output data is determined by the state of pin 5. You should note that these outputs can be tristated so the chip can be connected directly to a bus without the necessity of extra silicon to avoid bus contention problems.

Low-group input   The low-group input on pin 11 is designed to detect and accept the low-group signals. The waveform and amplitude requirements for this input are fairly tight, and you usually have to do some conditioning to the tones before presenting them to the 5102 or 5103. The 3525 takes care of this and can directly drive the inputs of the tone decoder without the need for additional components.

High-group input   The high-group input on pin 12 detects and accepts the high-group DTMF signals. Its input requirements are exactly the same as the low-group input on pin 11.

Other pins   The other pins are power (pin 1) and ground (pin 6). Even though this is a CMOS chip, both the 5102 and the 5103 are designed to run on 5 V. There's only a half-volt tolerance on these parts so you have to be careful about how you're going to power them—less than 5 V and the chip won't operate; more than 5 V and the chip won't ever operate. This is an especially important point to remember because the 3525 needs a minimum of 9 V.

*A slick trick*_____

One great thing about the 3525 that may not be immediately obvious is that since the chip is designed to separate two frequencies from a single mixed audio source, you can have it split out any two frequencies by changing the frequency of the crystal (and consequently the 3525's master clock). The only real restriction you'll have here is that the two frequencies you're looking for have to be within the range of the chip as it's defined by the IC's internal frequency circuitry.

One handy application is to have the 3525 do telephone dial-tone detection. The reason you can do this is that a standard dial tone is also just a combination of two frequencies—350 and 440 Hz. If you use a crystal of 1.758 MHz, you'll cause the 3525's low-group filter to pass all of the frequencies between 334 and 496 Hz. Since that includes both the dial-tone frequencies, you'll get a dial-tone indication at the low-group output and nothing at all at the high-group output. If you have a hard time finding one of the 1.758 crystals, you can divide the 3.58 crystal by 2, get a frequency of 1.79 MHz, and use that instead.

_____

Remember that these two chips, the 5102 and 5103, are pin-for-pin equivalents. The functional differences are so slight that if you have to make a choice, the basic qualifications of price and availability are the best guidelines. The bottom line in this case is simply:

Think with your wallet.

The needs we have in our project aren't making any great demands on the DTMF tone decoder. All we want to do is operate it in the most basic mode possible.

The schematic for the rest of the receiver—the demodulator and decoder—is shown in Fig. 10-11. If you think that an unusual number of parts are hanging off the 3525, remember that AMI made the chip extremely versatile, leaving it to the designer to add the passive components rather than burying them in the chip's substrate.

The rule is always that

Versatility breeds complexity.

If you want to be able to have lots of control, you're going to have to put the control components on the board yourself. That's the only way it can be. Lay the parts out on the breadboard as shown in Fig. 10-12.

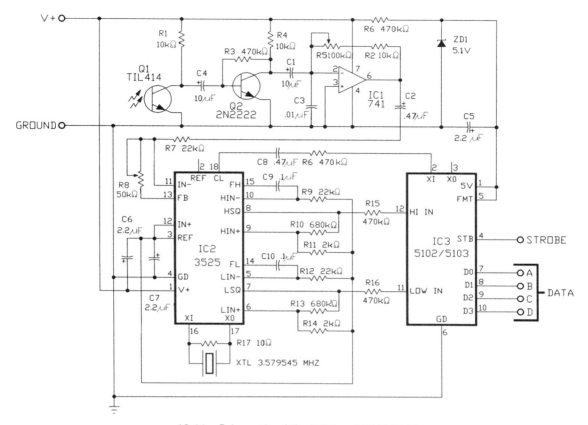

10-11 Schematic of the 3525 and 5102/5103.

You can see what most of the passive components are doing by using the 3525's block diagram as a reference. This may be a bit clearer if you look at Fig. 10-13. I've shown the components and the chip's block diagram together. You should be able to understand why most of the components are there, but let's run through the diagram.

Capacitor C2 decouples the input from the sensor/amplifier that detects the original received infrared and converts it into tone pairs. The potentiometer on the input op amp is a final trim adjustment for the gain, and you can use it to make sure the 3525 is being fed with the correct level. The data sheet for the chip (you do have one, don't you?) specifies a permissible range for the input level, but it's easier to set it on the fly by keeping an eye on the output of the whole receiver. Just use the potentiometer to trim the gain so that the whole receiver works properly;

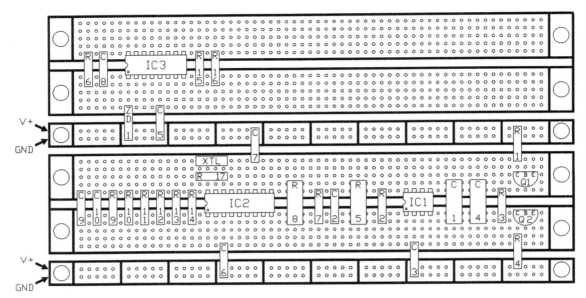

10-12   Placement diagram for the 3525, 5102, and 5103.

and once you've found that point, put a drop of nail polish on the potentiometer to lock it in place.

The two capacitors, C6 and C7, sitting from the power supply to the buffered reference voltage on pin 3, are to decouple the chip from the supply and to help bias the chip's internal op amp and comparators. At the output stages, the analog versions of the high- and low-group tones are capacitively coupled to the final comparators. All the remaining components are there to set up the gain of the output comparators. If you're not sure about any of these things, find out by getting any good book on op amps. I've even done some "Drawing Board" columns on op amps and comparators in *Radio Electronics* magazine. Check with your local library.

There are only two things to notice about the circuit that may be causing you a bit of confusion. The DTMF tone decoder—either the 5102 or the 5103—needs a crystal, but you'll remember that the chip can also be driven by a 3.58-MHz clock. That's how I've set things up here, and you'll notice in Fig. 10-11 that C8 is capacitively coupling the buffered version of the 3.58-MHz clock from the 3525 to the 5103. It's not much, but you do save a crystal. That's worth some board space and a couple of bucks as well.

The last important point to note in the schematic is the fact that the 5102 wants a 5-V supply and the 3525 has got to be powered by at least 9 V. As you can see in Fig. 10-11, I've used a zener diode, ZD1, to drop the voltage. However, as

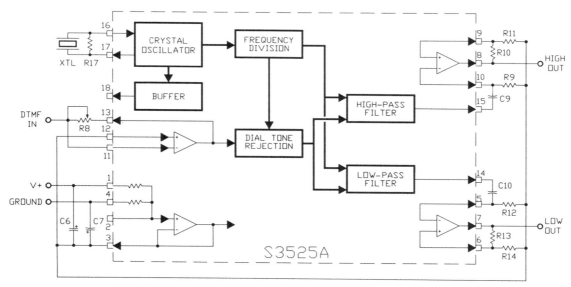

10-13 The 3525 block diagram with external circuit components.

shown in Fig. 10-14, you can also do the job with a standard regulator such as the 7805. The resistor on the 9-V line, R6, is there to protect the zener by limiting the current draw to something less than the zener's explosion level; but if you use the 7805, you can leave it out.

If you compare the two chips in the schematic, you'll see that the 5102 (or 5103) is more typical of what you ordinarily expect to see in dedicated telephone ICs. It was designed to do a particular job and it does it with a minimum—in this case absence—of external components. There's nothing electronically significant about this, but it is interesting to note because it again shows what we've already

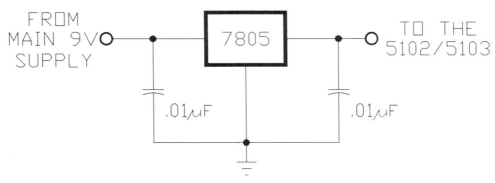

10-14 Using a 7805 to get 5 V from the 9-V supply.

seen—the 3525 is unique. It does a dedicated job but also gives the designer a lot of flexibility.

Once you have all the parts on the board, check to make sure you don't have a short between power and ground before throwing the switch to power up the board. This only takes a second. There's no way for me to know what kind of reading you're going to get, but I can tell you that it better not be zero. Checking for dead shorts is one of the most valuable habits you can have. I saw a friend of mine do it some years ago. It was one of those rare moments when you see someone do something and immediately say, "What a great idea!"

I've been doing it ever since, and you should get into the habit as well.

The receiver is a self-starting circuit, and it should be working as soon as you apply power. When you point the transmitter at the board and press a key, the infrared signal should be detected and the binary code should appear at the outputs of the 5103. We don't have anything on the outputs that makes it immediately obvious, but you can do some down-and-dirty detective work by snooping around the 5103 outputs with an LED or logic probe.

The 5103 latches the received data on its outputs so the binary code remains there even after the transmitter stops sending. If you want a hands-free detection method, put an LED and a current-limiting resistor on the strobe output of the 5103. This pin goes high when a valid signal is received at the chip's inputs and goes back low when the data disappears so you'll get a nice responsive indicator.

There are other easy ways to debug the circuit, but we'll get into them when we've finished the receiver.

Decoder

We've finally reached the last part of the block diagram we laid out in the beginning and, believe it or not, we're back to dealing with the familiar MSI stuff we all know and love. The final output of the circuit we've put together so far is the 4-bit binary data coming out of the DTMF decoder. What we have to do now is convert this into a 1-of-16 output. In this case, it's going to be an exact duplicate of the transmitter circuit that generated the binary code in the first place.

After everything we've been through so far, adding the decoder to the back end of the receiver is, in the words of Ralph Kramden, a "mere bag of shells." The schematic in Fig. 10-15 (although there's so little to it that calling it a schematic is a bit of overkill) uses the same 4514 we used in the transmitter. The data outputs of the 5102 or 5103 drive the corresponding inputs of the 4514, and the chip's latch is controlled by the strobe output.

Whenever valid data is detected by the receiver, the binary code is put on the inputs of the 4514, and the strobe output of the 5102/5103 goes high to latch the

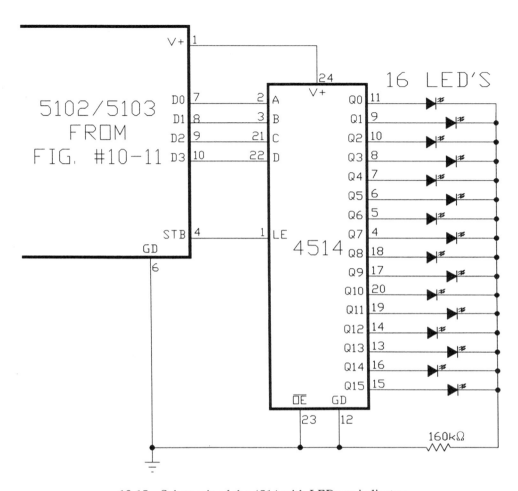

10-15 Schematic of the 4514 with LEDs as indicators.

data in the 4514. As a result, the key pressed on the transmitter will cause a corresponding LED to light at the back end of the receiver.

The minute you see one of the LEDs light up, you've got yourself a working remote control system. The outputs of the 4514 will control just about anything you can imagine; but if you've got plans for doing anything more than lighting an LED, you're going to need some extra stuff at the output.

Remember that the 4514 is a CMOS chip, and the drive capability is pretty limited. The receiver we built is just the controlling part of a complete system so you're going to have to put in some bench time to design the circuitry needed to safely connect the receiver to whatever stuff you want to control.

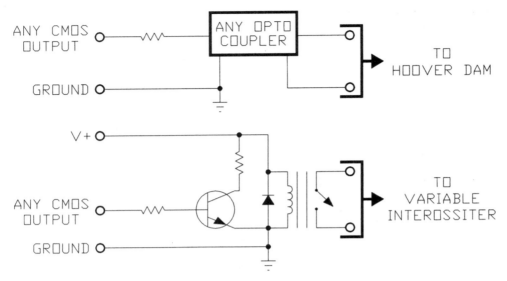

10-16 How to control external devices with CMOS outputs.

This isn't a major design job because there are lots of parts around that will let you hook up the receiver to anything else. These range from simple transistor drivers, to relays, to optical devices that can directly control a 120-Vac load. There's no way for me to know what you have in mind, but the circuit outlines in Fig. 10-16 should be able to point you in the right direction. These are just a few of the literally infinite number of ways you can get a CMOS-level output to safely drive everything from a lightbulb to the power modulation end of a variable interossiter—and if you manage to get one of them, give me a call. I'd like to see it.

Going further

Overcoming the inherent limited output drive of a CMOS part takes a few components and a few minutes of thought, but all of that stuff is more or less standard circuitry you can get from a book, a friend, or even (so I've been told) on the side of a bus. A much more serious problem with our remote control system—one that's potentially more limiting—is the fact that there are only 16 keys on the input of the transmitter and 16 outputs on the receiver. Now even though I know that a good number of the applications for our remote control system will be happy with just 16 codes, I'm just as sure that a good number of them won't be happy at all.

We've built a remote control system to activate one of the 16 outputs on the receiver. Expanding the number of outputs without adding any more keys on the transmitter is an interesting problem in design. It's also one you probably won't be able to answer by lifting a convenient, ready-to-go circuit from a data sheet or the pages of a general book on circuitry.

Some of the schemes that come to mind should be immediately thrown out the window, such as the one actually used for a while by one of the lesser-known scientists on Neptune—definitely not the renowned Klant Zorch. Without going into any of the nitty-gritty of either the circuit or the application (still regarded as top secret), two 4514s were addressed in parallel by the DTMF decoder and the active one was selected by throwing a switch. Whenever you wanted to use the remote control, you first had to go over to the receiver and throw the switch to select the 4514 you wanted to enable. An interesting idea, but only theoretically.

Although there are probably several ways to tackle a problem like this, I always like to go for the most logical one. If you can come up with something slicker, so much the better. The bottom line to increasing the number of outputs is that you have to add more outputs to the receiver. This may seem obvious—I hope it's obvious!—but the reason I mention it is only to emphasize the point that getting more outputs is a matter of control and decoding, not coming up with a scheme to add more codes.

The most obvious choice for an approach to designing something like this would be to make use of the 4514's output enable pin. If you go back over the schematics we've developed, you'll see that we haven't made any use of this pin at all. It's always been tied to ground to keep the chip permanently enabled.

In order to come up with a circuit scheme which would give us one active output from a choice of 32 or more possible outputs, one natural conclusion would be that we need this pin. After all, what we're trying to do is select not only a particular output, but a particular chip as well. Doing something on this order means we have to disable the outputs on all but one chip. We have to reserve a keycode which would be used to select the individual 4514 and then set things so the following code would select a particular output on that chip. A design like this would undoubtedly use loads of things like set/reset flip-flops, latches, decoders, and a certain amount of the always-needed logical glue.

Sounds pretty messy to me. In the words of a past president of the U.S.A., we could do it, but it would be wrong.

A particularly neat and unquestionably sneaky way to get the job done is shown in Fig. 10-17. It uses less silicon, a bare minimum of passive components, and is easily expandable to give you almost 10 times the output capacity—all without very much in the way of extra work, parts, or hassles of any kind. All you'd have to do is add a few wires and a couple of passive components. Pretty amazing.

Rather than reserve a keycode, we reserve one output on each of the 4514s you want to use in the receiver. The key behind the circuit is to leave all the 4514s enabled (just as we've always done) and talk to the 4514s one at a time by making use of the strobe output on the DTMF tone decoder.

When the receiver is first turned on, an RC pulse is generated to make sure the 4017 is reset and has a high on its first output. The store input controls (pin 1) of the 4514s are connected to and controlled by the outputs of the 4017. Since the first output is left floating, none of the 4514s have their internal latches open when the power is first turned on.

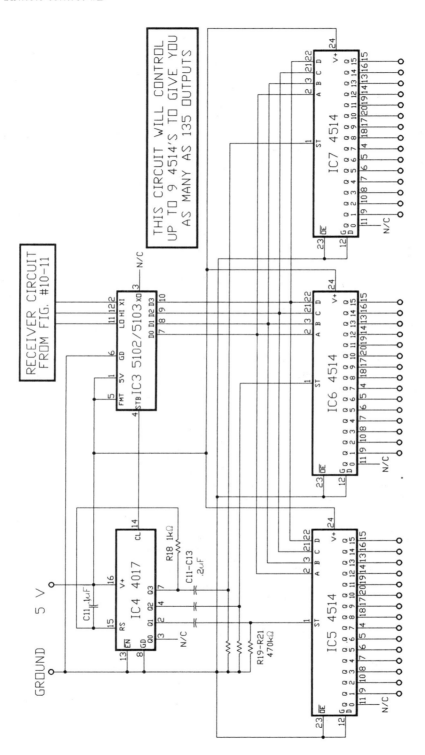

10-17   Schematic of the 4017-based output expansion circuit.

As soon as the receiver detects a DTMF tone pair and decodes it into 4-bit binary, the DTMF tone decoder puts a high on its strobe output and the binary code appears on its data outputs. The positive transition of the strobe output causes the 4017 to advance one count and put a high on its second output. This causes a fairly long high pulse to be sent to the store input of the first 4514, which will open the latch there. The result of all this is that the data coming from the DTMF decoder will be selecting an output on that, and only that, 4514.

When the next key is pressed on the transmitter, the strobe output of the DTMF decoder clocks the 4017 to the third output and generates an RC pulse to open the latch of the second 4514. This continues for as many 4514s as you want to have until the last one is reached. At that point, the high on the 4017's output not only generates the pulse for the store input of the 4514, but also causes the 4017 to reset and sit, once again, with a high on its first output.

There are several things to pay attention to and a few things to watch out for if you use this circuit. The first thing to note is that you have to use the strobe output of the DTMF decoder to control its data outputs as I've illustrated in Fig. 10-18. This is necessary because data appears on the outputs before the strobe output goes active. By configuring the 5102/5103 as shown in the schematic, the data outputs will be tristated until the strobe output goes high. We have to do this because we don't want any data on the output bus until the strobe has clocked the 4017 to the next 4514 in line.

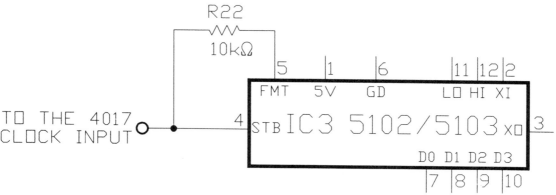

10-18 Using the 5103 strobe output to control enabling of the 5103 data outputs.

It may be easier to understand this if you study the timing diagram in Fig. 10-19. You can see that the strobe output causes the correct 4514 latch to stay closed until valid data is present on the bus. This means that the selected 4514 won't ever be able to put a high on the addressed output unless and until the 5103 has detected a valid DTMF pair. It's very important to understand the sequence of events because they're at the root of why the circuit is set up as shown in the sche-

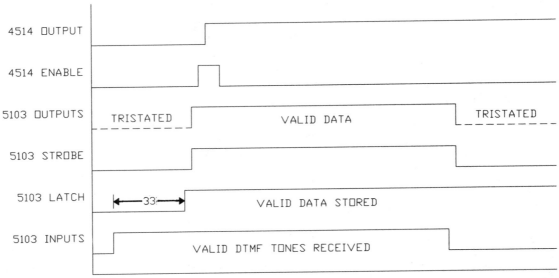

10-19   Timing diagram for the output expansion circuit.

matic. Any misunderstanding of this point will make the entire approach to the design seem mysterious. And don't forget that

You can't change something you don't understand.

Some things can be skipped over lightly, but the basic idea behind the design certainly isn't one of them.

You have a choice as to whether all 16 of the outputs on each 4514 should be used or whether one of them—preferably the first one on the chip, pin 11—should be left unconnected. This decision is necessary because the design of the circuit means you have to address all the 4514s whenever you use the remote control. Remember that the circuit will enable each 4514 in sequence until the 4017 is reset just after the last 4514 is addressed.

If you want to be able to address a particular 4514 without turning on anything, you have to have a null position; and the Q0 output is a good choice. If you'll always want the 4514s to turn something on, you can use all 16 of the outputs; but if you do that, you're going to have to put your "always on, default type" output on the Q0 output of the 4514.

The reason for this is that, no matter how carefully you arrange the timing in the circuit, there will always be a time when the 5103 outputs are tristated. This isn't a problem but it also means that the 4514 inputs will be left floating—and that is a problem. A floating CMOS input is always a major no-no.

To guard against that happening, the four resistors on the data line guarantee that the data lines will be all zeros when the outputs of the 5103 (or 5102) are tri-stating. This further means that the default 4514 output will always be the first one—the Q0 output on pin 11.

Now that we know all this, let's see how the multiple 4514 circuit has to be addressed by the transmitter.

Remember that the rest position (if you want to call it that) for the circuit is reached when the 4017 resets and its unconnected Q0 output is high. When the transmitter is used, it causes the 4017 to sequentially activate the store inputs of each of the 4514s in turn before the last one is reached and the circuit returns to its rest position. An addressing sequence by the transmitter, therefore, will involve exactly as many keystrokes as there are 4514s in the receiver to be addressed.

Although not strictly necessary, it's much better to sacrifice the first output on each 4514 and let it be a null, or not connected to anything, output. If there were three 4514s on the output of the receiver and you wanted to turn on the third output of the first 4514, none of the outputs on the second, and the sixth output on the last 4514, you would enter a 306 on the transmitter.

Remember that you have to enter a complete command sequence—in this case, three digits—every time you use the transmitter. If you enter fewer digits at the transmitter than there are 4514s on the receiver, you won't reset the 4017; and the next time that you use the transmitter, the first key you press won't be sent to the first 4514, and everything will be screwed up.

Now you can see why you're better off leaving the Q0 output of each 4514 unconnected and treating it as a null position. There's nothing special about that particular output, but it's much more intuitive to think of the 0 key on the keypad as the one that turns off the 4514 than, say, the 3 or 9 key. The 0 key was more logical for me, but the final choice is yours.

The last thing to watch out for are the RC pulses sent to the store inputs of the 4514s. I found the pulse generated by the resistor and capacitor combination to be adequate for the circuit. If you find the action of the circuit to be erratic in that the correct outputs on the 4514s don't always get turned on, try cleaning up the pulse. You can fool around with the values of the components to extend the high time or you can add some silicon to build a half monostable.

The schematic for this addition to the receiver is illustrated in Fig. 10-20. As you can see, a noninverting buffer is used to keep from inverting the pulse. I used a 4050 (I told you that I prefer to use CMOS stuff whenever I can), but any noninverting gate will get the job done just as well.

The complete schematic for the final DTMF receiver circuit is shown in Fig. 10-21, and the placement diagram is in Fig. 10-22. You probably have your own ideas about how to design the back end of the receiver, but it's a good idea to get things working before you start experimenting on your own. Remember that it's easier to modify than it is to create—at least in this case.

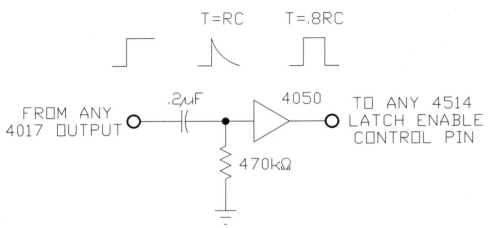

10-20   Adding a noninverting half monostable between outputs of the 4017 and latch controls of the 4514s.

If you're interested in exploring a really intriguing possibility, it occurs to me that there might be another, undoubtedly slicker, and certainly more interesting, way to add more outputs to the back end of the receiver. The method we just went through is pretty neat but this one, if it works out, would not only be neat, but definitely cool as well.

Before we get into this (and there isn't much to get into), let me emphasize that I haven't tried it yet. The theory is good but, as we all know,

Theory and practice are only related theoretically.

Paperwork always has to be backed up by boardwork.

The whole remote control, from the front of the transmitter to the back of the receiver, is based on DTMF tones. These are a set of very precisely defined frequencies, and the dedicated DTMF ICs we've been using have been designed to handle exactly those frequencies. To get that kind of precision, each DTMF chip uses a 3.58-MHz crystal to generate its master clock and then processes that clock through a set of internal dividers to get to the DTMF frequencies.

We've already seen that you can change the operating frequencies of the 3525 by changing the crystal frequency. Well, if you can do it to that chip, why not to the other DTMF chips as well?

The transmitter can have a manual switch or shift key to swap from one crystal to another. The receiver is a bit more complex. You can use some sort of flip-flop or set/reset latch arrangement to change the master clock frequency—a transmitted code would be decoded to do the electronic switching. This is one way to approach the problem, but it occurs to me that it's unnecessarily complex.

A much better approach would be to add another 3525/5103 chip pair to the

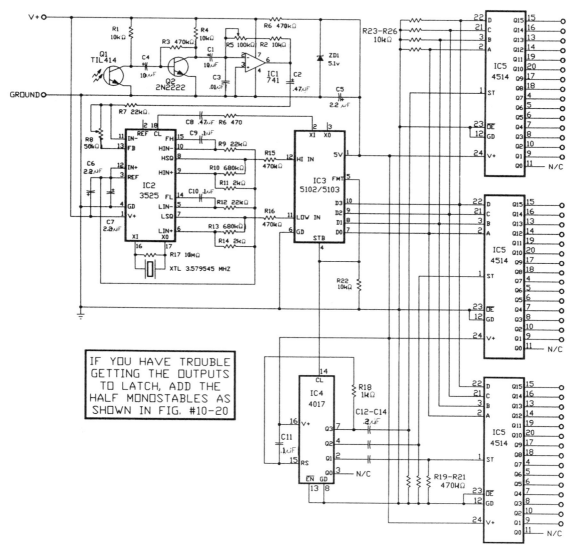

10-21 Complete receiver and decoder schematic.

receiver. One set would run off a 3.58-MHz crystal and handle the standard DTMF tones, while the other would run off a twin of the second crystal in the transmitter and handle the alternate tones. A second frequency of half the 3.58-MHz standard would be a good one to use because you could generate it with a simple divider circuit and eliminate the need for the second crystal. Not only that, but you'll probably have a hard time finding a 1.789 773-MHz crystal. It's easier to derive it from a 3.58-MHz master clock.

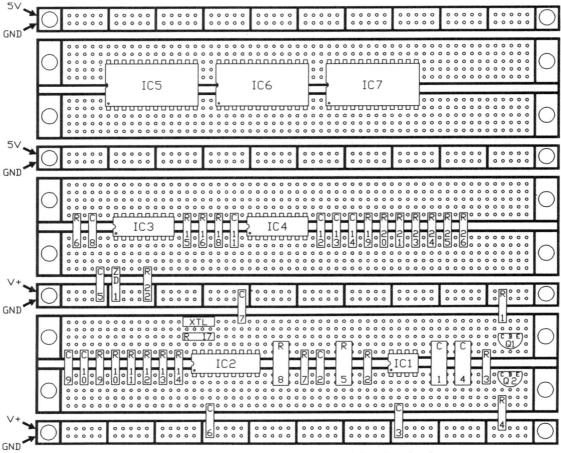

10-22   Placement diagram for receiver and decoder circuit.

Now remember that this is just a suggestion. I haven't tried it yet, but it's cool enough to pass along anyway. And it's probably not as easy as it seems because

Nothing simple in theory is simple in fact.

The more time you spend at the bench, the more you'll find out how true that really is.

Oops!

I've seen wiring mistakes in circuits that consisted of nothing more than a battery and a lightbulb so the first thing to look for should a problem crop up is a mechanical error of some kind. This can be anything from a wire in the wrong place, to a missing wire, to one of the components in backwards or not in at all. Bent pins are murder to detect.

When you're looking for signals on an IC, always check on the pins of the IC, not anywhere else. If you've bent one of the pins so it's folded under the chip, everything might look OK but the only way you can know for sure is to look for a signal on the pin itself.

Since we used dedicated ICs to build the demodulator and decoder, the circuit's not very complex; and if it's not working, the first thing to do is to carefully check the connections on the breadboard against the schematic. More than 80% of problems are caused by this sort of error. If everything's wired correctly and you're sure that all the components are properly placed in the proper orientation and have the proper value, the next troubleshooting step is to use the wham-bam method.

The first thing to do is to bypass the entire infrared section on both the transmitter and receiver. Connect the audio from the DTMF generator directly to the 3525 and try that. If nothing's still going on, use the line input of your stereo to listen to both the input and outputs of the 3525. You should hear a DTMF pair on the input pin of the 3525 and individual frequencies on both the analog and digital outputs.

Once you've verified that this is happening, you know that you've got a problem with the decoder section. If you're not getting tones on the 3525 outputs, especially the analog ones on pins 14 and 15, make sure the crystal is oscillating. You'll need a frequency counter or oscilloscope for that one—even Klant Zorch couldn't design a speaker with a frequency range that went up to 3.58 MHz.

The only areas that might cause you a problem with the 5103 are the crystal input (you are feeding the clock to pin 2, aren't you?) and the power supply (you are providing 5 V to the chip, right?). A 9-V supply to the 5103 causes a problem that can only be solved by a telephone call to your parts supplier.

If you're getting codes out of the 5103 but nothing is happening on the 4514, the first place to look is the power supply. The binary output from the 5103 is swinging between zero and 5 V, and that means you have to power the 4514 and the rest of the back end of the receiver with the same 5-V supply you're generating to feed the 5103.

You should also check the orientation of the LEDs on the outputs of the 4514. Remember that this chip has an active high output so the anode has to face the output of the IC. It's not always easy to know which leg of the LED is which even though there's a convention to make the anode leg a bit longer than the cathode. To give you an idea of exactly how strictly that convention is followed, I'm—right at this moment—holding six LEDs picked at random from the LED drawer of my parts box. Four of them have a long anode leg and two of them don't. I rest my case.

The only way to be sure of an LED's orientation is to use a battery and resistor and make the LED light up. A different but similar problem is found with capacitors. For reasons known only to the people who make capacitors, there's no convention whatsoever regarding the marking of the values. You've got dot code, color code, and a wide assortment of numbering schemes.

As you go through life, your head gets filled with various numbers that different agencies and organizations tell you are important to have on the tip of your tongue. I know my phone number, my checking account number, my social security number, various other phone numbers and customer numbers . . . I even know, for some reason or other, all 80 gazillion numbers on my driver's license.

But I refuse to memorize all the capacitor numbers. The best investment I ever made was a capacitor meter.

A last word

The DTMF-based remote control system we designed together is the most complete project in this book. You can use it as it is to do real things in the real world, or you can split off chunks of it for other things you might have in mind.

If you pull off a piece of the circuit to build, say, a circuit to show you the number you're dialing as you dial it, don't forget that a different circuit will have different circuit requirements. If you're lucky, you can plop something from this project into the middle of some other project and everything will work. But if you're like the rest of us, you'll undoubtedly have to do some modifications to get it all working in a new setting.

Modifying an existing circuit isn't difficult, but the only way you can do it is to first have a complete and total understanding of the circuit you want to modify. There's no better way to guarantee brain damage than trying to alter something without knowing exactly how it works in the first place, and why the circuit has been configured the way it is.

I'm the first person to disregard the labels that say No User Serviceable Parts or To Be Opened by Trained Personnel Only; but when I do that, I spend a long time looking and understanding—with the power off and my hands deep in my pockets. And I take notes—a lot of notes (not, however, with my hands in my pockets).

The design of the remote control system, particularly how we went about expanding the output, is a perfect example of what you can do if and when you understand how a circuit works. This is as true for the circuit that you design yourself as it is for one done by somebody else. Even though we didn't go through the formal steps of drawing up a list of criteria and a block diagram of the output expansion circuit, it's well worth your time to go back and do these steps anyway because any exercise in logical circuit analysis is valuable.

I gave you a suggestion of an alternate method for adding outputs. Because you have the circuit in front of you on the bench, you'll learn an awful lot by working your way through it. I haven't tried this method myself, and I can't guarantee that it will work; but I am sure the experience of trying it out on your own will be more valuable and teach you a lot more than any book you'll ever read. You really learn by thinking and doing, not reading and copying.

A slick trick

You can build neat-looking LED VU meters with an LM3915 and a few miscellaneous parts, but the one thing they're missing is the feature added to similar meters found in commercial equipment—a peak indicator. It only takes a few parts to build the circuitry necessary to add one of these to your own meters. This is a sample-and-hold circuit that will detect and remember the maximum level of a varying signal.

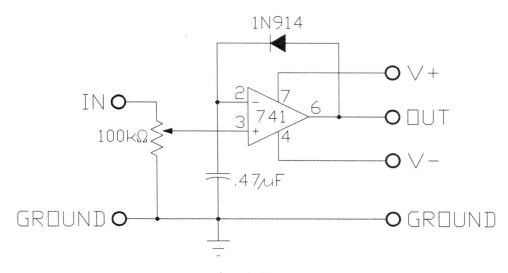

A peak detector.

The input voltage will forward bias the diode and cause the capacitor to charge. The charge will always be the same as the maximum voltage applied to the input of the op amp. The larger the capacitor, the longer it will be able to hold the peak. A value between 0.1 μF and 1 μF will provide storage times between a few seconds and a minute. If you feed the output to a low-impedance input, the loading will dramatically reduce the storage time.

11

So what now?
Out there on your own

Everything in life is attitude. I don't care how much you know about something or how much experience you have under your belt, when you get right down to it, if you don't believe you can do it—really and truly believe—then you can't do it. This may sound like the kind of stuff you hear on Sunday mornings at the ends of the dial, but that doesn't mean it's not true. It is.

I once lived in a basement apartment and along the ceiling in one of my closets was a whole bunch of plumbing. There were pipes going in every direction, and they were all joined together with elbows and tees. So one day I opened the closet and—you guessed it—there was a leak. I took everything out of the closet and looked carefully at the pipes. I could see where the leak was, but I couldn't figure out how it could possibly be fixed. If you loosened the pipe in one place, it would tighten somewhere else at the same time. No matter how long I thought about it— not too long, I admit—there was no way to take the whole pipe assembly apart.

When the plumber showed up, I hung around to watch him work because I wanted to see how he went about fixing the pipe. Knowing what I know now, some years later, I can tell you that this guy was an absolutely great plumber. He must have had a tenth-degree black belt in plumbing—real Olympic class—and he took care of the problem with a bare minimum of effort in an absolute minimum of time.

While he was working, I stood there and we talked about this and that (I mean, how much can you say about fixing a leaky pipe?); and the subject drifted off to the events of the day, recent movies, and a variety of other stuff.

He was the stupidest guy I ever met. But he was also a great plumber. As I stood there watching him doing his thing without working up a sweat, I thought to myself, this has got to be the stupidest person on the face of the Earth and here I am, a pretty smart guy with a college education. If he can do it, then so can I—I mean, how hard can it be?

Ever since then, I've done my own plumbing.

Everything is attitude. If you believe it can be done, then you have to believe that you can do it. It may take you longer, and you may go through some real brain damage to get there; but if you don't have any doubts, you don't have any problem. The minute you begin to doubt your own ability to do something, you might as well get out the yellow pages.

We've gone through a couple of designs together, but that's not what this book is really about. I wasn't trying to show you how to go about designing something in particular. I wanted you to learn how to design anything in general. It's all in the method—the secret's in how you approach the problem. What should stick in your mind when you finally put this book on the shelf is that any project you take on has to be managed in a systematic way or you'll never be able to see it through to a successful end. It's just not going to happen.

The backbone of boardwork is paperwork. If you don't wind up a project with a stack of notes and a few additions to your notebook, you've done something wrong. Every project there is, or ever will be, has to be taken through the same three stages of development—think about it, write it down, and finally wire it up. If you leave any one of these steps out, or even give one less attention than it needs, you'll wind up, sooner or later, paying a heavy price in time, money, and damaged brain cells.

And even if you do manage to complete the project, I'm willing to bet you my new stainless-steel butter baller with the ergonomic handle that, six months down the line, you'll have a hard time fixing it or explaining exactly how it works. Without good paperwork to back up the hardware, you're bound to wind up without a clue about why you did some of the things on the board.

Don't worry about producing the ultimate version of your project when you start working on it. Designing hardware means that, as you get further into the project, your mind is going to get totally and completely obsessed with it. The more time you spend working on it, the more intuitive everything's going to become. As you pump more and more stuff into your subconscious, it's going to do more and more work for you. And the more work you get it to do, the better your design is going to be. It can solve problems that leave you hanging over the bench with your tongue out.

It's not a bad idea to keep a pad and paper next to your bed. I can't tell you how many times I've suddenly sat up in bed, wide awake, at three in the morning with the answer to a problem that's been in my mind for days. This is a great feeling, but it only happens when you really get into the design. The most unusual thing about it is that the answer you get from your subconscious is always miles away from anything you were consciously thinking about. The more abstract the problem you're working on, the more likely you are to get something useful from your subconscious because it generally provides a new way of seeing the overall problem, rather than nitty-gritty details of the design such as component values and so on.

But even if you have the world's most active subconscious, it's not going to be able to do a thing for you if you can't feed it the right information. You can't design in a vacuum, and that means you need a reference library. A major part of your library is your own notebook but you need more—much more. Whenever you add a particular IC to the board, you're really adding a bunch of circuitry designed by someone else. Remember that there's a lot going on inside even the simplest integrated circuit. You've got to know everything there is to know about a particular IC if you decide to make it a player in your project. That means databooks—lots and lots of databooks.

Every single semiconductor manufacturer in the universe spends a lot of time and money to create and support a huge library devoted to only one thing—listing, explaining, and showing how to use all the chips and other things the company produces. The larger the manufacturer, the more extensive the paperwork. The people who write this stuff may not win any prizes for literature, but they know everything there is to know about the hardware. And they want to tell you everything they know.

Appendix A lists the names, addresses, and phone numbers of the major semiconductor manufacturers. You should get in touch with every one of them to get a list of the databooks and other literature they have available. The larger companies—Intel, Motorola, and so on—publish literature guides you can get just for the price of a phone call or letter. These guides are a complete index of everything they publish, and you should have them in your library.

The first place you should turn when you have a particular project in mind is the list of available literature. Aside from the databook containing the parts you want to use, you'll probably find application notes listed as well. These are definitely things that you want to get your hands on. They're real-world, practical discussions of how the parts should be used, and they usually include examples of working circuits.

How well you're going to do at the bench depends on how well you prepare the bench before you start working. Have a clean surface and all the books you need within arm's reach. As you get into the design, you'll find yourself spending as much time reading as wiring, so it's a good idea to keep your notebook handy. You never know when bright ideas are going to pop into your brain. The more experience you get with them, the more you'll appreciate just how quickly these bright ideas come and, more important, how quickly they go.

These flashes of insight travel at 837 times the speed of light and don't spend a lot of time hanging around the forefront of your brain. If you don't write them down as soon as they show up, they'll be gone forever.

Designing circuits is a combination of logical thought, adequate literature, good habits, and momentary inspiration. All these things have to be present if you want to maximize your chance of winding up with a working version of what was originally a good idea. You can learn how to be systematic, you can buy the databooks, you can even develop good habits; but the only way to get those occasional hits of subconscious stimulation is to work on your own, put in the time, and get the experience.

The more you do the more you'll be able to do, and time spent on your own is the most valuable kind of time there is. Working from a book is a good way to get started, but you're not really going to turn into a designer unless and until you do alone everything we've been doing together.

So close the door, tune the radio to some sort of mindless trash, stock up on junk food, and take your brain out of cruisamatic.

Get to work.

Appendix A
Semiconductor manufacturers
Where to get literature

Every single one of the semiconductor manufacturers listed here has a complete line of databooks, and you should get material from every one of them. The material they have available is, without question, one of the most valuable sources of information and help available to a designer, regardless of what the project is. Call or write to every one of them.

There's a charge for their literature, so the extent of your library depends on the extent of your bankbook. At the least, you should get their publication lists and choose the books you want to have around as permanent additions to your reference library. They're not the kind of books you'll want to read before going to bed; but when you're in the middle of a project, they can provide you with that one piece of information you desperately need.

And you don't ever throw out old databooks just because you get a newer edition.

Mostek Corporation
1215 West Crosby Rd.
Carrolton, TX 75006
(214) 466-6000

National Semiconductor
2900 Semiconductor Dr.
Santa Clara, CA 95052-8090
(408) 721-5000

Gould AMI Semiconductors
3800 Homestead Rd.
Santa Clara, CA 95051
(408) 246-0330

Klant Zorch Electronics
83 Whotsomanfdhr Way
Mxdrtfjkl, NP 059683
(987) 555-5634

Motorola Semiconductor Products
5005 East McDowell Rd.
Phoenix, AZ 85008
(602) 244-6900

Signetics Company
811 East Arques Ave.
Sunnyvale, CA 94088-3409
(408) 991-2000

Intel Literature Department
3065 Bowers Ave.
Santa Clara, CA 95051
(800) 548-4725

Hitachi America, Limited
2210 O'Toole Ave.
San Jose, CA 95131
(408) 435-8000

Sprague Integrated Circuits
115 Northeast Cutoff
Worcester, MA 01606
(617) 853-5000

NEC Electronics
1 Natick Executive Park
Natick, MA 01760
(617) 655-8833

RCA Corporation
Borton Landing Rd.
Moorestown, NJ 08057
(609) 338-3000

Texas Instruments
13500 North Central Expwy.
Dallas, TX 75265
(800) 232-3200

Although I've built up a fairly large collection of databooks over the years (and always get new ones), I obviously didn't use all of them as references during the writing of this book.

The ones I did use for one thing or another are the following:

Gould AMI MOS Products Catalog
Mostek Microelectronic Databook
Signetics CMOS 4000B IC Family Databook
Klant Zorch Unobtainium Databook
National Semiconductor Telecommunications Databook
National Semiconductor Linear Databook

These are really good books to own—get copies of them for yourself.

Before ending this discussion on literature, I've got to make sure that you understand that however great databooks are, they're not the only source of good information. The consumer magazines like *Radio Electronics* and trade publications like *EE Designer* always have good stuff in them; and they're the place to turn for circuit tips, notes on what's new in the industry, and other terrific tidbits.

And don't forget the U.S. Government Printing Office. It's the largest publisher in the world.

You can never read enough about a wide enough variety of subjects because you never know what stuff is going to be published and where it might show up. You can find out anything about anything. I've been convinced for years that the notion of classified and/or top secret research is a joke because it's always true that the material finds its way into print somewhere.

The most significant circuitry is based on novel ideas, not novel parts. You may have some trouble getting 3 pounds of enriched uranium or an ingot of unobtainium oxide, but don't get the idea that everything that can be done with standard parts has already been done by someone else. Nobody thinks the way you think, so nobody will do what you'll do.

Think about it.

Even the X-ray laser stuff used in the less-than-impressive Star Wars program is in print. You're not going to find detailed circuit diagrams, but you will find all the theory and research results.

Interesting, huh?

Appendix B
Setting up a bench
What you have to have

As electronic designs get more sophisticated, so does the minimum equipment you need on the bench. I've picked the projects in this book carefully so that you can build them with a minimum amount of test equipment. But if you want to do heavy-duty stuff, you're going to need more than just a battery, logic probe, and multimeter.

The most basic piece of test equipment you can own is one that you'll find yourself using more than anything else—of course, I'm talking about an oscilloscope. These used to cost as much as the national debt of certain South American countries, but the prices have dropped unbelievably in the last couple of years.

At the last IEEE show I went to, I saw 30-MHz dual-trace scopes with all the usual goodies selling for well under 400 bucks retail, and that means around 300 bucks mail order. The pages of *Radio Electronics* are loaded with ads for inexpensive scopes, and even though I haven't personally tried all of them, the ones I have seen are well-designed instruments.

The oscilloscope market has undergone the same sort of changes that took place in the stereo business. Once upon a time you could clearly identify some stereos as good and some as, well, let's just say less than good.

Today's crop of stereos are basically terrific, and all of the price differences are based on extra features—the bells and whistles. You can't find a stereo in today's market that doesn't deliver sound that a normal person would call more than acceptable. You can pay through the nose for sets that will reproduce the mating calls of bees and the hunting sounds of bats, but find me a pair of speakers that can handle that.

There's no getting around the fact that you still need traditional equipment like logic probes, test clips, multimeters, and a reasonable power supply; but the way things are now, you also need a scope. It's the fundamental piece of test equipment. Recent wall paintings found in Egypt show that some of the ancient architects used a primitive, wind-powered scope to design the pyramids.

There are absolutely no rules to tell you how your bench should be set up. This is as personal a matter as the projects you design on it, and the only guideline you should follow is to put together something that works for you. If you're a neat freak, you can put things inside plastic boxes. If you believe, as I do, that a neat bench is the sign of a narrow mind, you can have things as cluttered as you want. It's up to you. Whatever works.

I would no more tell you how to set up your workbench than I would tell you what to do on it. Remember that starting with an idea and winding up with a completed project is an intensely personal activity that depends as much on individual creativity as it does on the laws of electronics.

Your workbench is your business and, with the exception of having an oscilloscope sitting on it, you should set it up exactly the way you want to do the work you want to do.

It's your place and, ultimately, your business.

Index